Cooking with Low-cost PROTEINS

Cooking with Low-cost PROTEINS

Betty Ann Clamp

ARCO PUBLISHING COMPANY, INC.
219 Park Avenue South, New York, N.Y. 10003

Published by Arco Publishing Company, Inc.
219 Park Avenue South, New York, N.Y. 10003

Copyright © 1976 by Betty Ann Clamp

All rights reserved. No part of this book may be reproduced, by any means, without permission in writing from the publisher, except by a reviewer who wishes to quote brief excerpts in connection with a review in a magazine or newspaper.

Library of Congress Catalog Card Number 74-19770
ISBN 0-668-03638-9

Printed in the United States of America

This book is dedicated to my two boys,
Charlie and Kurt,
and to all children throughout the world.
I hope that in some small way this book
will help to make happier and healthier children,
wherever they may be.

ACKNOWLEDGMENTS

To my many wonderful friends, students, and family members who lent support, encouragement, ideas, and recipes, I wish to extend a most sincere thank-you. A special thank-you is given to Eleanor Ogle, whose special efforts helped make this book possible.

Thanks also to the Quong Hop Company of South San Francisco, California, for the samples of tofu, along with recipes, that I worked with.

Last but not least, I wish to thank my brother, Ralph Muehlenbruch, for doing the photographic work.

Contents

List of Tables	8
Introduction	9
What Protein Is	11
Foods Which Supply Protein	13
Why We Need Protein	18
Protein's Relationship to a Total Meal Plan	22
Recipes	24
Soybeans	26
Tofu	39
Textured Vegetable Protein (TVP)	54
Legumes	68
Grains, Nuts, and Seeds	82
Breads	96
Eggs and Cheese	116
Meat Stretchers	132
Fish	152
References	167
Index	169

List of Tables

Table I	General Protein Content of Selected Foods	16
Table II	Amino Acid Content of Certain Foods	16
Table III	Complementary List of Certain Foods	17
Table IV	Recommended Daily Dietary Allowances for Protein	20
Table V	Daily Guide to Basic Four Food Groups	22
Table VI	Comparison of Soybeans, Soybean Products, and Other Legumes	29
Table VII	Nutrient Composition of Selected Grains, Nuts, and Seeds	83
Table VIII	Nutrient Content of Selected Flours and Grain Products	98
Table IX	Nutrient Content of Selected Foods	117
Table X	Nutrient Composition of Some Cooked Meats	133
Table XI	Nutrient Content of Selected Fish	154

Introduction

Shortages, boycotts, strikes, and high prices have sharpened public interest in utilizing various protein sources. A great many people, including myself, have come to the realization that adjustments and changes may have to be made in our food choices, and that this may require the use of smaller portions of meat with an increased use of vegetable protein sources such as legumes, grains, tofu, and textured vegetable proteins.

I first realized the need for and interest in using a greater variety of protein foods in my nutrition class. During a discussion of proteins, I prepared a spaghetti sauce made with ground, cooked soybeans rather than meat. The student response was quite enthusiastic, and my superior, Dr. Robert Niederholzer, encouraged me to author and teach a most successful course entitled "Cooking With Low-Cost Proteins." This book is an outgrowth of the class.

Today, many Americans actually eat more protein than they need. However, tomorrow may be a different story due to increasing prices and an increasing world population competing for the food which is produced. Adequate protein intake could become a problem for most Americans. If the fundamental concepts presented in this book are understood, one should be able to make wise protein choices. Also, the recipes given are varied enough so that almost everyone should find something to his liking.

The reader must also realize that this book deals primarily with one nutrient: protein. However, other nutrients—vitamins, minerals, carbohydrates, and fats (especially polyunsaturated)—in addition to water and some fiber are also necessary for life and good health. Eating a large variety of foods from all the food groups will help to insure an adequate supply of nutrients. Therefore, we must not rely on only a few foods to meet our protein and other nutrient needs, and should expand our tastes to include many and, possibly, new foods.

What Protein Is

The word protein was derived from a Greek word meaning *to come first*. It is of prime importance to all living organisms, whether plant or animal, and is found in all living tissues. It is a complex substance containing carbon, hydrogen, oxygen, and nitrogen. Many proteins contain additional elements such as iodine, iron, phosphorus, and sulfur. Since the human body cannot utilize free nitrogen, we must obtain it from the protein in the food we eat.

Complex protein structures are actually built from smaller units called amino acids. If amino acids occurred free in the food we eat, they are all that we would need to obtain our nitrogen. However, in nature they occur linked together in protein molecules, and only through the process of digestion are they made available to the body.

Twenty-two amino acids commonly occur in nature and some twenty amino acids are known to be building blocks of the proteins in the human body. Given the right building materials, some of these amino acids can be manufactured by the body. However, there are eight that the human adult cannot manufacture. For this reason they have been termed *Essential Amino Acids*. They cannot be manufactured from elements present within the body, and must be ingested in completed form as part of the food consumed.

ESSENTIAL AMINO ACIDS

Isoleucine	Phenylalanine	Histidine [1]
Leucine	Threonine	Arginine [2]
Lysine	Tryptophan	
Methionine	Valine	

[1] Additional requirement of children.
[2] May not be manufactured fast enough to meet the body's needs under all conditions.

The amino acids which the body can manufacture are termed *Non-essential Amino Acids*, not because they are unimportant, but because the body can produce them when it has enough of the necessary elements present at the time they are needed.

NON-ESSENTIAL AMINO ACIDS

Alanine	Glycine
Aspartic Acid	Hydroxyproline
Cysteine [3]	Proline
Cystine [3]	Serine
Glutamic Acid	Tyrosine

[3] Sulfur containing amino acids.

Both the essential and non-essential amino acids are linked together to form the proteins found in the foods we eat.

Foods Which Supply Protein

Since all living matter contains some protein, the plant and animal foods we eat all contain protein. However, the amount contained in foods varies. Most fruits contain one percent or less protein and therefore are not really considered a protein source. Meats, fish, poultry, and cheese contain roughly 25 percent protein and are considered to be excellent protein sources.

Percentages tell the most concentrated sources of a nutrient, but the amount eaten is what really matters. For example, cooked soybeans are about 11 percent protein and provide approximately 18 grams of protein in a 170 gram (1 cup) serving—one-fourth to one-third of an adult's daily requirement. A slice of bread contains only 2 grams of protein, but in a sandwich the bread contributes 4 grams of protein. See Table I for the protein content of some selected foods.

It is not enough for a food to contain a given amount of protein. The protein must be digestible and must also contain sufficient quantities of the essential amino acids. For example, the protein in corn is quite deficient in lysine and tryptophan, and gelatin lacks tryptophan altogether. The proteins in these foods will not support life by themselves. The amino acid or acids present in insufficient quantity are called the limiting acids in the food; for example, lysine and tryptophan are limiting amino acids in corn.

Egg protein, along with human milk protein, contains

the essential amino acids in the best amounts or best pattern for human use. Therefore, it is considered to be of the best or highest quality and is used as a reference standard to which other proteins are compared. Except for gelatin, all animal proteins rank fairly high in protein quality. Animal proteins are very good for body building, even if eaten alone, because the essential amino acid content and pattern is much like that of the proteins in our own bodies.

On the whole, plant proteins contain insufficient amounts of one or more of the essential amino acids, and the amino acid pattern is unlike that of the proteins in our bodies. Consequently, the plant proteins (except for soybeans) rate low on the quality scale. When eaten alone, they are not very good for body building.

The fact that a protein may be low in quality does not mean it has little value. For example, corn is not adequate to sustain life by itself. However, it will supply some other necessary amino acids when eaten in combination with a food that is high in tryptophan and lysine. Beans are such a food, but rice is not. Table II gives a general description of the amino acid strengths and weaknesses of certain foods.

A knowledge of the strengths and weaknesses of the amino acid content of foods provides the necessary background for combining foods to provide high quality protein in which all the essential amino acids are present. One generalization to keep in mind when meal planning is that grains are weak in lysine but strong in methionine and that legumes are strong in lysine but weak in methionine, and that grains are therefore a good combination with legumes. Examples of grain-legume mixtures are baked beans with brown bread, lentils with rice, and refried beans with tortillas. Peanut butter and bread are a grain-legume mixture, but the amino acid strengths and weaknesses of the two are somewhat alike and a glass of milk would improve the protein quality of this combination a great deal. Eating a combination of foods with the same weakness (such as rice and corn or rice and bread) only accentuates the lysine deficiency of the grains.

Animal foods can be used to supplement a weakness in a plant food. Examples of such a combination are grated cheese on beans, cheese and macaroni, and casserole dishes using small amounts of meat. By using a mixture of vegetable proteins and animal protein, the more expensive animal protein is extended or stretched and the quality of the protein is improved. Table III lists a number of foods that complement each other in ways that have been described.

Although it is possible to achieve high quality protein through choosing plant proteins carefully, most nutritionists recommend that some animal protein be included in the diet, especially for a child. Not only will one be sure of getting enough high quality protein, but animal foods are the only ones that contain vitamin B_{12}. I often prepare black-eyed peas and cornbread for dinner, and when I do, I make sure that my children drink milk. The animal protein could also come from other sources, such as a meaty ham hock.

It is to our advantage, nutritionally and economically, to use more plant foods in our diets. Animals are more expensive to produce and difficult to keep. Most plants can be easily grown in the backyard or even in window boxes.

The recipes section of this book provides recipes which are either high quality vegetable protein combinations or high quality animal-vegetable protein combinations.

FOODS WHICH SUPPLY PROTEIN

Table I
General Protein Content of Selected Foods

Food	Percent Protein	Average Serving	Protein in Average Serving
Most meat, fish, poultry, cheeses	20–28	3 oz. (84 g.)	17–23 g.
Most nuts and seeds	15–24	2 tablespoons (18 g.)	3–4 g.
Eggs	13	1 egg (50 g.)	6.5 g.
Breads	8–10	1 slice (25 g.)	2–2.5 g.
Legumes—cooked	5–11	1 cup (190 g.)	10–21 g.
Pasta—cooked	4	1 cup (140 g.)	6 g.
Cake	3–5	1 piece (54 g.)	2–3 g.
Milk	3.5	1 cup (244 g.)	9 g.
Deep green vegetables	2–3	½ cup (128 g.)	2.5–4 g.
Corn, wheat, and rice—cooked	2–3	1 cup (205–245 g.)	3–4 g.
Most fruit	½–2	whole fruit	½–1 g.

Adapted from: Deutsch, Ronald M., *The Family Guide to Better Food and Better Health,* Meredith Corporation. 1971; *Composition of Foods,* Agriculture Handbook No. 8, Agricultural Research Service, United States Department of Agriculture, 1963; and *Nutritive Value of Foods,* Home and Garden Bulletin No. 72, United States Department of Agriculture, 1971.

Table II [1]
Amino Acid Content of Certain Foods

Food	Strength	Weakness
Seafood, meats, poultry	Lysine	None
Dairy products	Lysine	None
Legumes	Lysine	Methionine
Nuts and seeds	Methionine	Lysine
Grains, cereals, and cereal products	Methionine	Lysine
Fresh vegetables	Lysine (O.K.)	Methionine

Table III [1]
Complementary List of Certain Foods

Food	Combine with
Seafood, meat, poultry	Grains and cereal products, legumes, nuts, seeds
Eggs	Grains and cereal products, legumes, nuts, seeds
Dairy products	Grains and cereal products, legumes, nuts, seeds
Legumes [2]	Grains, nuts, seeds, animal foods
Seeds and nuts	Fresh vegetables
Grains and cereal products [3]	Dairy products, other animal foods, legumes, fresh vegetables especially greens
Mushrooms	Beans, green peas, brussels sprouts, broccoli, cauliflower
Yeast	Grains and cereal products

[1] Adapted from Lappe, Frances Moore, *Diet for a Small Planet*, Ballantine Books, Inc., 1971.
[2] Some good examples of legumes are soybeans, navy beans, black-eyed peas, lentils, lima beans, pinto beans, peas, and red beans.
[3] Grains are wheat, corn, rice, oats, barley, and rye. Some common products are flour, breads, and cereals.

Why We Need Protein

Through the processes of digestion and absorption, the amino acids in food protein are made available to our cells for production of body proteins and for their functions.

The most obvious need for protein is seen in the structure of the human body. Every cell contains some kind of protein. A great need for high quality protein exists during childhood when tissue growth is occurring. In fact, protein deficiency can retard a child's growth. The formation of tissues in the fetus increases a pregnant woman's need for high quality protein. The production of milk on the part of a breast feeding mother also increases her need for protein. The formation of hemoglobin, the substance in the blood which carries the oxygen to the cells, is most dependent upon the presence of protein in the diet and protein need may be increased following a loss of blood.

When growth stops, protein is still necessary for the maintenance, replacement, or repair of damaged cells. The cell replacement we most commonly see is that of hair, skin, and nails. Once maturity is reached, the protein need remains fairly constant except during stress conditions such as fever, illness, hemorrhage, surgery, pregnancy, and lactation. Protein requirement is increased somewhat under these conditions.

Protein participates in a number of regulatory functions. Protein is one factor involved in preventing the blood from becoming too acidic or too alkaline. It helps to maintain the proper balance of water between the cell and the blood. The protein molecule stays in the blood and exerts the pressure necessary to pull the water out of the cell into the blood. When there is insufficient protein, water accumulates in the tissues (a condition known as edema).

Protein is also necessary for the production of enzymes, some coenzymes, hormones, and antibodies. Enzymes make it possible for the reactions within the body to occur and with great speed. Enzymes sometimes require a coenzyme (a vitamin plus protein as part of their structure) to work. Hormones such as adrenalin, insulin, thyroxine, etc. have protein as part of their composition, and these hormones help to regulate certain activities within the body. Antibodies help to fight infection, and the body's ability to fight infection is decreased in protein deficiency.

Any surplus tryptophan (an essential amino acid) not required in body protein synthesis can be converted to niacin (vitamin B_3).[1]

Protein can also be utilized for energy: it provides four calories [2] per gram. In fact, protein that is left over from the previously described needs is converted to energy and any protein not used for energy is stored as fat. All in all, protein is a more expensive source of energy than fats or carbohydrates.

Now you may be asking "Well, how much protein is needed to adequately fulfill these functions?" Since in many cases the average American's diet consists of more meat than he really needs,[3] it is doubtful that many Americans suffer from protein deficiency at this time. Problems

---

[1] Sixty mg of tryptophan = one mg of niacin.
[2] These are really kilocalories.
[3] About a three ounce serving two times a day is sufficient.

arise when diets consist primarily of grain foods, as in the case in low-income areas. If someone understands the principles which have been explained and uses this knowledge, he should not suffer from protein deficiency even though rich protein sources, such as meat, become less accessible.

The Food and Nutrition Board of the National Academy of Sciences-National Research Council has set up a guideline for us to determine if we are meeting our protein needs as well as other nutrient needs. These guidelines are known as the Recommended Daily Dietary Allowances. The guidelines for protein intake are in Table IV.

Table IV
Recommended Daily Dietary Allowances for Protein [1]

Persons	Years	Protein in Grams
Infants	0.0–0.5	13.2
	0.5–1.0	18
Children	1–3	23
	4–6	30
	7–10	36
Males	11–14	44
	15–18	54
	19–22	52
	23–50	56
	51+	56
Females	11–14	44
	15–18	48
	19–22	46
	23–50	46
	51+	46
	pregnant	76
	lactating (nursing)	66

[1] Adapted from *Recommended Dietary Allowances,* Revised 1974. Published by the National Academy of Sciences-National Research Council, Washington, D.C., 20418.

The allowance for protein is determined by age, sex, size, pregnancy or lactation, and amount of muscle tissue. The protein requirement in proportion to body size is the greatest during periods of rapid growth when new tissues are being formed. It decreases as the rate of growth slows. A larger person will have a somewhat greater protein need than a smaller person. Protein requirements increase following surgery or hemorrhage and during pregnancy and lactation.

Exercise and activity are not factors in determining protein requirement unless the diet is deficient in calories from fats or carbohydrates.

Protein's Relationship to a Total Meal Plan

Protein should not be the sole or even principal concern in meal planning. The major concern should be to eat a wide variety and good balance of foods so that we receive all the nutrients provided by foods. There is no food that provides all the nutrients we need. By eating a variety of foods, we should get a good balance of amino acids as well.

Food guides have been developed to help us in our food selection so that we obtain all the necessary nutrients. The Basic Four shown in Table V is a very commonly used food guide in the United States and some other countries.

Table V
Daily Guide to Basic Four Food Groups

Group	Servings per Day (cups)	
Milk—cheese, ice cream, and other milk-made foods.	Adults	2 or more
	Children	3 or more
	Teenagers	4 or more
Meat—meat, fish, poultry, eggs, cheese, dry beans and peas, nuts.	2 or more (3oz or 84g)	
Vegetables and Fruits—include dark green or yellow vegetables, citrus fruit or tomatoes.	2 or more vegetables 2 or more fruits	

Bread—cereal, grains such as   4 or more
  rice, wheat, corn, oats, and
  grain products (enriched or
  whole grain).

In addition, other foods may be used as needed to give **complete** and satisfying meals. Such foods include butter, salad dressings, jellies, jams, and condiments.

Most of our protein is provided by the Milk and Meat groups. Note that two servings of milk products and two servings of meat, fish, poultry, legumes, or nuts are recommended. This is sufficient to supply the greatest portion of our protein requirement. The protein supplied in the recommended serving of dark, leafy green vegetables and grain products provides the remainder.

Recipes

For the adventurous cook who is eager to make the changeover, the recipes are presented in metric in addition to avoirdupois.

The metric measurements were determined in the following manner for the purpose of this book.

Liquid or volume measurements were determined through the use of equivalent tables. When uniform measuring equipment becomes available, it may be necessary to round off the quantities to the nearest measurable volume. For example 118 ml will probably be measured as 125 ml and 237 ml as 250 ml. At the present Corning is promoting a Pyrex® measuring cup and Foley has put out a plastic measuring cup with markings at 25 ml intervals or gradations. If you own such a cup you may wish to measure according to the following chart.

Milliliters Given in Recipe	Pyrex® or Foley Cup
59 ml	50 ml or 0.5 dl
78 ml	75 ml or 0.75 dl
118 ml	125 ml or 1.25 dl
158 ml	150 ml or 1.5 dl
177 ml	175 ml or 1.75 dl
237 ml	250 ml or 2.5 dl or ¼ l
296 ml	300 ml or 3.0 dl
315 ml	325 ml or 3.25 dl
355 ml	350 ml or 3.5 dl
395 ml	400 ml or 4.0 dl
414 ml	425 ml or 4.25 dl
474 ml	500 ml or 5.0 dl or ½ l

Weight measurements, except for teaspoons and most tablespoons, were derived either from equivalent tables or by actually weighing the food on a diet scale, which weighed in 10 gram intervals up to 450 grams. If a canned food was used, the metric weight was taken from the level—if one was given. Otherwise, the pounds and/or ounces were converted to the metric equivalent. For quantities greater than one pound or 450 grams, a large kitchen scale, which measured in 50 gram increments, was used. You may find it necessary to round the amounts given to the nearest measureable weight, probably in either a 10 or 25 gram increment.

Dry measures of teaspoons and tablespoons were not converted as we will probably continue to use the tablespoon, equal to 15 ml, and the teaspoon, equal to 5 ml. Such is the case with a set of spoons I've seen from Japan, only the spoons are labeled as 15 ml, 5 ml. 2.5 ml, etc. Other countries on metric also use spoons which are comparable for small measurements. See the following:

UNITED STATES		SPAIN	
1 cup = 236.6 ml		1 taza	= 200 ml
16 tablespoons			= 16 cucharadas
1 tablespoon = 3 teaspoons		1 cucharada	= 3 cucharaditas
GERMANY		SWEDEN	
1 Suppenteller = 250 ml		1 tekopp	= 250 ml
16 Esslöffeln		1 kaffekopp	= 150 ml
1 Essölffel 3 Teelöffeln			= 10 matsked
		1 matsked	= 3 tesked

Each recipe section includes general information, cooking tips, and nutrition information comparing, nutrients often found on the nutrition label of many foods. These nutrients are important but it is essential to remember that there are additional nutrients, not included, which are necessary for life and good health.

Soybeans

GENERAL INFORMATION

In some countries, notably the United States, the soybean has been used primarily for animal food, and animals have thrived on such feed. The soybean is a perfectly good food for human consumption however, and interest in eating the soybean and its products has been growing. Hopefully, it will soon become easier to find soybeans and soybean products in local supermarkets.

Basically, there are two types of soybeans: the field type which is used in the production of oil and commercial flour, and the vegetable type which is larger and milder in flavor. The vegetable type is good eating when green and fresh or after it has been dried and then cooked. After thorough cooking it is tender but remains firm in texture, not soft or mealy.

There are numerous soybean products which are beneficial and fun to use: flour, grits, tofu, textured vegetable protein extenders or analogs, and sprouts are some examples. Some of these products are described in more detail in subsequent chapters.

NUTRITIVE VALUE

The soybean's protein content is greater than that of other dried beans and peas. In 100 grams of dried soybeans, cooked, there are 11 grams of protein. This yields approximately 18 grams of protein in a one cup (about 170 g) serving. A one cup serving of other dried beans, cooked, yields about 13 grams of protein. The amino acid content and pattern of soybean is also closer to that in the human body than that of other beans and peas. In fact, the protein quality is fairly close to that of meat. The essential amino acid methionine is somewhat deficient. The addition of methionine to soy protein improves the quality, and quality can also be improved by combining soy protein with a grain product or with a food of animal origin.

The mature soybean is a dependable source of a number of vitamins and minerals, including calcium, phosphorus, iron, riboflavin, and thiamin. The fresh green soybean also provides a good amount of vitamin A, much of which is lost during the drying process. The fresh bean and the sprout contain a little vitamin C.

The soybean contains less carbohydrate and more fat than other beans, but it is still relatively low in fat and contains no cholesterol. The fat it does contain is polyunsaturated, and it can therefore be useful in cholesterol-restricted diets. It is also rich in lecithin, the value of which is under controversy. See Table VI for an approximate analysis of soybeans, soy products, and other legumes.

COOKING TIPS

1. Do not cook with soybeans that have been treated for seed purposes.
2. Acquire a food grinder of some kind.
3. Soak before cooking. Allow 4 cups (940 ml) water to each cup (approximately 175 g) of dry soybeans. Boil for 2 minutes and then let stand for 1 hour or let stand overnight in the refrigerator.
4. Cook the beans after soaking. You can use the soaking water but may need to add additional

water. Add 1 teaspoon (6 g) of salt for each cup (175 g) of dry soybeans used. Simmer until tender, about 3 hours. Remember that soybeans do not get really soft.

5. To reduce foaming, add 2 teaspoons (10 ml) oil or meat drippings.
6. Cook a large potful of beans at one time and freeze in 1 or 2 cup (170 g or 340 g) portions for later use. Freezing seems to soften the texture a little.
7. Some people claim that sprouting the bean slightly before cooking helps to reduce flatulence, commonly known as gas. Others feel that soaking and long slow cooking will help to eliminate the problem.
8. Cooking, baking, or any mild heat treatment renders the protein in the soybean or other beans and peas more digestible. One reason for this is the presence of an inhibitor in the legume which interferes with the protein digesting enzyme trypsin. Heat destroys this inhibitor.
9. Use the soybean in recipes as you would use other beans. Grind it and use it in loaf or patty recipes. It is also good to use the ground soybean mixed with hamburger to extend the hamburger.
10. Cook in a pressure cooker to save time and energy. Do not fill the pressure cooker over one-third full of beans and water. Cook at 15 pounds pressure for about 30 minutes.
11. After oil extraction, the soybean is often ground into soy flour or grits. The grits are just like the defatted flour, but more coarsely ground. Add a little of these products to cooked or baked foods and increase the protein value.
12. Full fat soy flour contains polyunsaturated fat and is the kind of soy flour most commonly available

Table VI
Comparison of Soybeans, Soybean Products, and Other Legumes, 100 g [1]

Food	Calories	Protein g	Fat g	Carbohydrate g	Calcium mg	Iron mg	Vitamin A I.U.	Thiamin mg	Riboflavin mg	Niacin mg	Ascorbic Acid mg
Black-eyed peas, cooked	76	5.1	.3	13.8	17	1.3	10	.16	.04	.4	—
Most dry beans, cooked	118	7.8	.6	21.2	50	2.7	6	.14	.07	.7	0
Lentils, cooked	106	7.8	trace	19.3	25	2.1	20	.07	.06	.6	0
Soybeans, green, cooked	118	9.8	5.1	10.1	60	2.5	660	.31	.13	1.2	17.
Soybeans, dry, cooked	130	11.	5.7	10.8	73	2.7	30	.21	.09	.6	0
Soybean sprouts, cooked	38	5.3	1.4	3.7	43	.7	80	.16	.15	.7	13.
Soybean flour, full fat	421	36.7	20.3	30.4	199	8.4	110	.85	.31	2.1	0
Soybean grits, defatted	326	47.0	.9	38.1	265	11.1	40	1.09	.34	2.6	0
Tofu	72	7.8	4.2	2.4	128	1.9	0	.06	.03	.1	0
Textured Soy Protein	280†	52.†	1.0†	31.5†	220†	10.0†	—*	0.3*	0.6*	16.0*	trace*

[1] Values taken from *Composition of Foods*, United States Department of Agriculture, Handbook 8, 1963.
† *Protein-ettes* (Pamphlet), The Creamette Company, Minneapolis, Minn., 1974.
* Hamdy, M. M., "Nutritional Aspects in Textured Soy Proteins," *Journal of American Oil Chemists' Society, 51*: 85A, January 1974.

on store shelves. It should be stored in an airtight container in a cool place to retard spoilage.
13. Look for dried soybeans in the dried beans and peas section of your store or in the Oriental foods section. Ask your grocer to carry them if he doesn't. You can also buy them in specialty food stores and health food stores.

Fresh Green Soybeans in Butter

2 cups soybeans, shelled (340 g)
1 cup water, boiling (235 ml)
½ teaspoon salt
1 tablespoon butter

To shell soybeans (the pod is not edible) cover beans with boiling water and let stand for 5 minutes. Then break pod crosswise and squeeze out the beans. Cook the beans in the boiling salted water for 10 to 20 minutes. The beans will be tender but not soft and mealy. Season with butter and serve. Serves 4.

Serve with Spanish rice, spinach salad, and sherbet.

Golden Delight Salad

2 cup soybeans, cooked (340 g)
1 can corn (8¾ oz/244 g)
2 medium stalks celery, diced
2 green onions, chopped
¼ cup green pepper, diced (30 g)
2 tablespoons pimiento, diced

Combine ingredients and mix with the following dressing to taste. Serves 4 to 6.

Dressing

⅓ cup sugar (90 g)
½ teaspoon dry mustard

1 teaspoon salt
2 tablespoons flour
1 egg, beaten
½ cup vinegar (118 ml)
½ cup water (118 ml)
1 tablespoon butter

Mix dry ingredients. Beat dry mixture into the egg. Heat vinegar, water, and butter. Remove from heat and gradually add egg mixture, stirring fast. Put back to cook, stirring and constantly, for 2 or 3 minutes until smooth and thick. This dressing is also good for other salads and coleslaw.

Serve for lunch or a warm weather supper with bread or rolls and Yogurt Pie (page 119).

Soybean Salad

2 tablespoons oil (30 ml)
1 tablespoon lemon juice or wine vinegar (15 ml)
salt and pepper to taste
2 cups soybeans, cooked (340 g)
2 green onions, chopped
1 dill pickle, chopped
4 tomatoes, cut into wedges
2 cucumbers, sliced
4 large lettuce leaves

Combine oil, lemon juice, salt, and pepper. Add soybeans and toss lightly. Add green onions and dill pickle and mix. Serve on lettuce leaves with tomato wedges and sliced cucumber. Serves 4.

Serve with Parmesan Biscuit Bake (page 102) to complement the soybean.

Variation: Add 2 cups (340 g) cooked rice or macaroni.

Soybean Minestrone Soup

This recipe was the favorite soybean recipe in my classes.

- 1 tablespoon oil (15 ml)
- 2 cups soybeans, cooked (340 g)
- 1 can tomatoes, pureed (454 g)
- 1 can tomato sauce (227 g)
- 2 carrots, diced
- 1 large stalk celery, diced
- ½ onion (medium), diced fine
- 1 clove garlic, minced
- ½ bunch of spinach, chopped, or 1 package frozen spinach
- ½ cup macaroni, enriched or whole wheat (60 g)
- 1 teaspoon Italian seasoning
- 1 beef bouillon cube

Parmesan cheese, grated

Combine all ingredients except cheese in a large soup pot and simmer until tender and well blended. Garnish with grated Parmesan cheese when serving. Serves 6–8.

Soybean Gumbo Soup

- 2 cups soybeans, cooked (340 g)
- 2 tablespoons oil or margarine (30 ml)
- 1 onion, diced
- 1 clove garlic, minced
- 1 can okra with red peppers (1 lb/454 g)
- 3 stalks celery with tops, diced
- 1 cup brown rice, uncooked (190 g)

1 bay leaf
1 teaspoon thyme
3 chicken bouillon cubes *
2 tablespoons imitation bacon bits
1 can tomatoes (1 lb/454 g)
3 cans tomato sauce (8 oz/227 g)
8 cups water (1880 ml or 1.9 l) *

Combine all ingredients and cook for 2 hours. Serves 12.

Serve with Quick Cheese Bread (page 103), ice cream, and Granola cookies. The combination of the soybeans with the rice improves the protein quality of both.

Soybean Chile

4 cups soybeans, cooked (680 g)
1 small can of green chiles, diced (4 oz/113 g)
1 small can of tomato sauce (8 oz/227 g)
1 teaspoon onion flakes
1 clove garlic, minced
½ teaspoon salt
1 cup water (235 ml)

Combine all ingredients in a saucepan and cook slowly for about 1 hour. Serves 4–6.

Serve with cornbread, cheese bread, tortillas, or rice (see BREADS or GRAINS).

---

* Eight (8) cups of chicken broth may be substituted for the bouillon and water.

Soybean Spaghetti Sauce

1 tablespoon oil (15 ml)
½ onion, diced
1 clove garlic, minced
4 sliced mushrooms, optional
2 cups cooked soybeans—ground or mashed (340 g)
1 can stewed tomatoes (1 lb/454 g)
1 small can tomato sauce (8 oz/227 g)
1 can tomato soup (10¾ oz/305 g)
½ cup ketchup (118 ml)
½ teaspoon oregano, ground
½ teaspoon marjoram, ground
½ teaspoon rosemary
½ teaspoon sugar, optional
pinch of cinnamon
salt and pepper to taste
1 cup grated longhorn or mild Cheddar cheese (100 g)
Parmesan cheese, grated

Sauté the soybeans, onion, garlic, and mushrooms in the oil. Then add the tomatoes, tomato sauce, tomato soup, ketchup, and seasonings. Simmer for about 3 hours and add water if necessary. Stir in the cheese just before serving. Serve sauce over cooked spaghetti. Whole wheat spaghetti is very good. Top with Parmesan cheese. Serves 6.

Soybean Spaghetti Skillet

2 cups soybeans, cooked (340 g)
2 tablespoons oil (30 ml)
½ medium onion, diced

1 small can pimiento, chopped
1 can stewed tomatoes (1 lb/454 g)
23 medium, pitted black olives, sliced
1 teaspoon garlic salt
1 cup Cheddar cheese, grated (100 g)
½ pound spaghetti, cooked (227 g)

While spaghetti is cooking, sauté soybeans and onions in oil. Then add soybeans, tomatoes, pimiento, garlic salt, and olives. Cook about 20 minutes. Then add spaghetti and cheese. When the cheese has melted it is ready to serve. Serves 4–6.

The spaghetti and cheese complement the soybeans.

Soybean Tomato Cheese Casserole

2 cups soybeans, cooked (340 g)
1 package frozen corn (280 g)
1 can stewed or plain tomatoes (1 lb/454 g)
1 teaspoon sugar (optional)
1 teaspoon seasoned salt
½ teaspoon monosodium glutamate (optional)
¾ cup bread crumbs, buttered (40 g)
2 tablespoons wheat germ
½ cup cheese, grated (50 g)
pinch of pepper
paprika

In a buttered baking dish, arrange layers of corn and soybeans. Mix tomatoes and seasonings and pour over bean mixture. Top with bread crumbs, wheat germ, and cheese. Sprinkle top with paprika. Bake uncovered for 30 minutes at 375° F (190° C). Serves 6.

The soybeans are complemented by the corn, bread crumbs, and wheat germ.

Baked Soybeans

3 cups cooked soybeans (460 g)
½ onion, diced
½ teaspoon salt or to taste
2 strips bacon or 2 pieces of lean Canadian bacon, diced
2 tablespoons molasses (30 ml)
½ teaspoon dry mustard
1 tablespoon ketchup
½ cup water (118 ml)

Combine all ingredients in a covered casserole. Bake at 325° F (163° C) for about 2 hours. Serves 4.

Serve with some type of bread or rolls (see BREADS section) to complement the soybean protein.

Soy Flake Timbales

3 green onions, chopped
1 large tomato, chopped
2 tablespoons oil (30 ml)
2 eggs, beaten
1 cup milk (235 ml)
1 cup soy flakes, cooked (300 g)
2 teaspoons soy sauce (10 ml)
2 cups bread crumbs, preferably whole wheat (100 g)

Sauté onions and tomatoes in oil for 5 minutes. Add the vegetables to the milk and eggs. Add the bread crumbs, soy flakes, and soy sauce. Put into individual timbale or custard cups or large casserole. Bake at 350° F (177° C) for 25 minutes or until set. Serves 6.

Preparation of Soy Flakes

1 cup soy flakes, dry (105 g)
2 cups water (470 ml)
1 teaspoon lime juice or lemon juice (5 ml)

Bring the water to a boil. Add the soy flakes and reduce heat. Simmer covered for 1 hour. Stir in the skins which rise to the top and add lime juice. Makes about 1½ cups (300 g).

Soybean Patties

  2 cups cooked soybeans, ground (340 g)
1¼ cups water (294 ml)
1⅓ cups oatmeal (115 g)
  ½ medium onion, chopped
  ½ teaspoon oregano
  2 tablespoons soy sauce (30 ml)
  ½ teaspoon salt
  2 tablespoons oil (30 ml)

Mix soybeans, water, oatmeal, onions, seasonings, and oil. Shape into patties and fry in lightly oiled skillet until brown. Serve with tomato sauce given below. Makes 8 patties. You may also put inside hamburger buns and garnish like hamburger.

Tomato Sauce

½ medium onion, chopped
2 teaspoons oil (10 ml)
1 can tomato soup (305 g)
⅔ cup water (153 ml)
½ teaspoon Italian seasoning
½ teaspoon salt

Lightly sauté onion in oil. Add tomato soup, water, and seasonings. Cook slowly for about 15 minutes.

Soy Nuts

2 cups soybeans, dry (305 g)
salt
oil or shortening for deep fat frying

Soak soybeans in water to cover overnight. Drain, then set out on a towel for about an hour. Fry in deep fat at 350° F (177° C) for 10 minutes. Drain on paper towels or a brown paper bag. Salt and serve for a nutritious, crunchy, and tasty snack that won't promote cavities.

Tofu

GENERAL INFORMATION

The Quong Hop Company of South San Francisco, California, gives the following definition for tofu:

"Tofu is the coagulated protein of the soybean, extracted from the whole bean by soaking, grinding, cooking, and filtering the beans with water to produce a milk which is then processed into a curd in much the same way cottage cheese is derived from cow's milk. Its appearance is that of a light cheese, its consistency that of a firm custard, and its taste very plain, making it very adaptable to any dish or seasoning you choose to use." [1]

NUTRITIVE VALUE

The protein content of tofu varies from 7 to 20 percent, depending on the water content. The softer tofus contain more water and a lower concentration of protein. Some deep fried tofus are firmer and contain less water; hence a larger percentage of protein. One-quarter pound (113 g) of the medium-firm tofu provides about 8 grams of protein.

The quality of the protein is like that of the soybean from which it is made. The Quong Hop Company claims that its tofu contains substantial amounts of vitamin A,

---

[1] *Products Fact Sheet,* The Quong Hop Company, South San Francisco, California.

B complex, E, and K, and that it is rich in lecithin, and free of cholesterol. Depending on the coagulation or curdling process, tofu can be a good source of calcium.

COOKING TIPS

1. Cut the tofu into cubes and sprinkle it with soy sauce and eat it. This is the way some of my Japanese friends eat tofu.
2. Mix it or use it with almost any food. Since it is bland, tofu readily absorbs the flavor of other foods. Tofu makes an especially good meat stretcher and mixes well with hamburger. It adds protein to vegetable dishes.
3. Use the tofu in recipes in which the water that cooks out of the tofu won't make the recipe too moist.
4. Tofu is usually sold in one-pound (454 g) cartons. Sometimes these large cakes are cut into four equal pieces or smaller cakes. If your tofu has not been cut into four cakes, you should do so for use in the recipes in this section.
5. If the tofu needs to keep its shape in a recipe, pan fry it a little before using it in the recipe.
6. Tofu should be well drained before use. An easy way to drain it is to let it rest in a colander.

7. Keep tofu under constant refrigeration. Slit the top along one edge and allow the water to drain out. Flush the contents under cold water, fill to cover the tofu, and store in your refrigerator. Repeat this process daily and the tofu will keep for a week or more. It's okay until it begins to smell sour.
8. Fresh tofu is usually found in the fresh produce section or cheese section of a grocery store. You can also purchase it canned or packaged as an instant powder.
9. Leftover tofu can be frozen and used later in soups.
10. Japanese-style tofu is soft, whereas the Chinese-style is medium firm.

Tofu Scrambled Eggs

4 eggs
1 green onion with stem, chopped fine
1 tomato, chopped
2 cakes firm tofu, diced in ¼-inch (0.6 cm) cubes (225 g)
2 tablespoons soy sauce (30 ml)
Garlic salt to taste
2 tablespoons margarine or oil (30 ml)

Combine all the ingredients except oil. Heat oil or margarine in a frying pan and cook as you would scrambled eggs. Serves 4.

Serve with whole wheat toast and jelly.

Tofu Tomato Soup

    2 cups chicken broth or water (470 ml)
    2 cakes firm tofu, diced in 1-inch (2.5 cm) cubes (225 g)
    1 tablespoon soy sauce (15 ml)
1½ teaspoons salt
    2 teaspoons sesame oil (10 ml)
    2 tablespoons soybean oil or salad oil (30 ml)
    2 tomatoes, diced, with skin and seeds removed if so desired
    2 green onions, chopped fine
pinch of monosodium glutamate (optional)

Bring the chicken broth or water to a boil and add the tofu. Then add salt and soy sauce and bring to a boil again. Add the remainder of the ingredients and simmer for 5 minutes. Serves 4.

Serve with crackers or Cheese Sourdough French Bread (page 114) to complement the tofu. This soup would be good as a luncheon dish or served before dinner.

Tofu Cucumber Soup

4 cups chicken broth, may be canned (940 ml)
1 cucumber, peeled and diced
2 cakes of tofu, crumbled or diced (227 g)
1 teaspoon soy sauce (5 ml)
1 egg, beaten
2 green onions, chopped

Add the tofu, cucumber, and soy sauce to the chicken broth. Simmer over low heat 10 to 15 minutes or longer. Add the beaten egg to the soup and swirl quickly, to give an "egg flower" effect. Garnish with green onions. Serves 4–6.

Serve with crackers or rolls. This makes a good luncheon dish or a good way to start dinner.

Tofu and Vegetables

½ lb. fresh vegetables, thinly sliced (225 g)
2 cakes tofu, diced or cubed (225 g)
2 tablespoons yellow onions, chopped
½ teaspoon honey or brown sugar
½ teaspoon water (2.5 ml)
2 teaspoons soy sauce (10 ml)
oil for stir frying or sauteeing

Fry vegetables quickly over high heat. Remove vegetables, wipe pan, add more oil, throw in onions and add the tofu, seasoning with soy sauce, a dash of salt, and honey. Turn a few times, add water, and simmer covered for 10 minutes. Add the partially cooked vegetables and simmer 5 minutes more. Serves 4–6.

Serve with teriyaki (page 138) and steamed rice.

Tofu Oatmeal Patties

½ container tofu (2 cakes) well drained (225 g)
¾ cup quick cooking oatmeal (65 g)
2 green onions with tops, chopped
1 carrot, grated
½ teaspoon salt

Mix all ingredients well and shape into patties. To begin with, heat 1 tablespoon margarine or oil and fry patties until golden brown. Add more fat as needed. Top with soy sauce, sour cream, or unflavored yogurt. The tomato sauce recipe given with Chile Rellenos on page 120 is very good to top the Tofu patties. Serves 4–6.

The oatmeal complements the tofu protein so it is not necessary to use animal protein. However, the addition of 1 or 2 eggs may improve flavor and protein quality and will help to bind the patties together more.

Variations: Add ½ diced green pepper or add a handful of bean sprouts in place of onions.

Tofu Fish Croquettes

1 small can of tuna, salmon, mackerel (6½oz/183 g)
1 green onion, chopped
1 small stalk of celery, chopped fine
1 cake tofu (115 g)
white sauce
1 cup regular or toasted wheat germ or bread crumbs (50 g)
oil for frying in deep fat

Mix fish, onion, celery, tofu, and white sauce. Chill thoroughly. Shape into patties or cylinders. Roll in the wheat germ or bread crumbs and fry until lightly browned. You may want to dip in beaten egg before rolling in crumbs. Serves 4–6.

White Sauce

  2 tablespoons butter
  2 tablespoons flour
pinch of salt and pepper
½ cup milk (118 g)

Melt the butter. Add the flour, salt, and pepper and cook until bubbly. Stir in the milk and cook until thick.

Tofu Hamburgers

1 lb. ground beef (454 g)
2 cakes tofu (225 g)
1 green onion, chopped
1 clove garlic, minced

1 tablespoon soy sauce (15 ml)
1 tablespoon flour

Mix all ingredients in a bowl and shape into patties. Fry or broil. Makes 6 patties.

Serve with hamburger buns and condiments.

Tofu Meatballs

1 lb. ground beef or other ground meats such as turkey, chicken, pork, lamb (454 g)
4 cakes hard tofu, Chinese style (1 lb/454 g)
2 green onions, chopped
2 tablespoons soy sauce (30 ml)
1 teaspoon salt
2 tablespoons brown sugar
2 tablespoons sesame oil (30 ml)
pepper to taste

Mix all ingredients well and form into 1-inch (2.5 cm) meatballs. Bake at 400° F (205° C) for 20 minutes. Serve hot, either plain or with sweet and sour sauce. Stick with a toothpick or cocktail pick and eat as an hors d'oeuvre. Let sit in sweet and sour sauce (see Sweet and Sour Tofu recipe following) for several hours.

Serve with brown gravy over rice or noodles. The protein present in the rice or noodles will be complemented. Use this recipe for hamburgers or meatloaf as well. The meat improves the protein value of the tofu and the tofu stretches the meat.

Sweet and Sour Tofu

1 package hard tofu, fried
1 yellow onion
2 carrots
2 small green bell peppers
1 can pineapple chunks (20 oz/560 g)
sweet and sour sauce (canned, bottled, or recipe below)

Cut the vegetables into thin pieces 1½ inches (3.7 cm) long. Cut the cakes of fried tofu into four equal triangles. Stir-fry the vegetables for 2 to 3 minutes. Then add the tofu and turn gently a few times. Pour the sweet and sour sauce over the tofu and vegetables and top with the pineapple chunks. Cover and simmer 3–5 minutes. Serves 4–6.

Serve with rice, which together with the tofu will provide a complete protein.

Sweet and Sour Sauce

¾ cup brown sugar (120 g)
2 tablespoons soy sauce (30 ml)
½ cup cider vinegar (118 g)
2 tablespoons flour mixed with some juice from the pineapples

Mix all sauce ingredients in a saucepan and add the juice from the can of pineapples. Cook until slightly thickened.

Tofu Beef or Pork

1 lb. boneless beef or pork, sliced thin (454 g)
2 teaspoons soy sauce (10 ml)
2 teaspoons brown sugar or honey (10 ml)

1 tablespoon cornstarch or flour
2 green onions, cut in 1½-inch (3.7 cm) lengths
1 clove garlic, mashed
1 carrot, sliced thin
½ small head regular or Chinese cabbage
2 cakes tofu, sliced in ½-inch (1.2 cm) slices
¾ cup water (177 ml)
oil for stir frying

Marinate meat with soy sauce and sugar for a few minutes. Then roll in cornstarch or flour. Fry meat, green onions, and garlic until brown. Remove and clean pan. Add more oil, then stir fry carrot and cabbage. Remove and clean pan. Add more oil and tofu and turn a few times, sprinkling with a dash of soy sauce. Add water, meat, and vegetables and cook for 10 minutes. Serves 4–6.

Serve with rice.

Tofu Pork Hash Patties

2 cakes tofu (225 g)
1 lb. ground pork (454 g)
½ lb. shrimp, chopped (225 g)
1 small can water chestnuts, chopped (6 oz/116 g)
1 small carrot, grated
2 green onions, chopped fine
1 teaspoon salt
pepper to taste

Mix all ingredients, shape into patties, and pan fry or broil. Serves 8.

Serve with fried rice, a vegetable like broccoli, and a fruit salad.

Tofu Patties

1 cake tofu, well drained (120 g)
½ cup ham or Spam, chopped (80 g)
1 egg
1 to 2 green onions, chopped fine
salt and pepper to taste
1 tablespoon sesame seeds (optional)

Mix all ingredients, form into patties, and pan fry until browned. Serves 4.

Serve between hamburger buns or whatever else you may like.

Tofu Stuffed Manicotti

2 cakes tofu (225 g)
1 pound lean hamburger (454 g)
½ onion, chopped fine
1 clove garlic
spaghetti sauce or marinara sauce
½ green pepper, chopped fine
4 oz. Mozzarella cheese, grated (110 g)
8 to 12 manicotti shells, precooked for 8 minutes

Mix tofu and hamburger. Divide mixture in half and brown one-half with onion and garlic. When brown, add your favorite ingredients for spaghetti sauce or add marinara sauce. Simmer while preparing the remainder of the recipe.

Brown remainder of tofu and meat mixture with green pepper. When brown be sure to drain any grease, then stir in the grated cheese. When melted, stuff the manicotti with the mixture. Arrange in greased casserole or baking pan. Pour the spaghetti sauce over it. Bake about 30 minutes at 350° F (177° C). Serves 4–6.

Tofu Lasagne

 1 pound ground meat (454 g)
½ onion, diced
 1 clove garlic, minced
 1 small can tomato sauce (8 oz./227 g)
 1 small can tomato soup (8 oz/227 g)
 1 can tomatoes, pureed in blender (1 lb/454 g)
¼ teaspoon *each* of rosemary, basil, oregano, and pepper
 1 teaspoon salt
pinch of cinnamon
 1 cup water (235 ml)
½ package lasagne noodles, cooked (10 oz/140 g)
 2 cakes tofu, drained and crumbled
½ pound Mozzarella or Monterey Jack cheese, sliced (225 g)

Brown the ground meat with the onion and garlic. Add the tomato soup, tomato sauce, tomatoes, spices, and seasonings. Add the water and simmer for at least 3 hours. Add more water if necessary but keep the sauce thick. This sauce freezes well for future use.

Line the bottom of a greased rectangular baking dish with some of the noodles. Then layer with crumbled tofu and meat sauce. If you have some Parmesan cheese handy, it is good to sprinkle 1 tablespoon of Parmesan cheese on each layer. Top with meat sauce and slices of Mozzarella or Monterey Jack cheese. Bake at 350° F (177° C) for 20 to 30 minutes. The lasagne should be bubbly and the cheese melted. Serves 6–8.

Variation: Omit the ground meat and double the amount of tofu and cheese which are used.

Tofu Macaroni Salad

½ package macaroni (½ lb/225 g)
2 cakes hard tofu, cut in small cubes (225 g)
2 sweet or dill pickles, diced
½ medium onion, diced
½ cup mayonnaise or salad dressing (118 ml)
2 hard cooked eggs
⅓ cup pickle juice (78 ml)
salt and pepper to taste
lettuce leaves

Cook the macaroni according to directions on package. After macaroni is cooked and drained, add tofu, pickles, onion, mayonnaise, pickle juice, salt and pepper. Slice eggs and place on top. Serve on lettuce leaves and garnish with olives and paprika. Serves 6–8.

Tofu Fish Salad or Sandwich Spread

1 can fish such as tuna, mackerel, salmon, drained (6½ oz/183 g)
1 cake tofu, drained and crumbled (115 g)
¼ cup mayonnaise (59 ml)
1 pickle, chopped
1 small stalk of celery, diced fine
salt and pepper to taste

Combine all ingredients thoroughly. Make a sandwich, preferably using whole wheat bread. Spread on crackers for a snack. Stuff tomatoes for a main dish salad or stuff cherry tomatoes for snacks.

Tofu Tempura

2 cakes tofu, hard (225 g)
1 carrot, chopped fine
1 cup green beans, chopped fine (140 g)
2 green onions chopped fine

¼ cup sesame seeds (25 g)
1 cup cracker crumbs (60 g)

Mix these ingredients and dip by tablespoonfuls into the following batter or use a commercial mix for batter. Deep fry in hot peanut oil.

Tempura Batter
1 egg
1½ cups water (353 ml)
2 teaspoons baking powder
2 cups flour (250 g)

Tofu Stuffed Cabbage Rolls

This recipe was adapted from one given to me by a German neighbor.

8 large cabbage leaves
½ lb. ground meat (225 g)
2 cakes tofu (½ lb/225 g)
1 teaspoon salt
¼ teaspoon pepper
¼ cup onion, chopped (40 g)
   pinch of nutmeg
¼ teaspoon sage
1 egg, slightly beaten
2 tablespoons butter

Cook cabbage leaves in boiling, salted water, a few minutes to soften; drain. Combine the remaining ingredients and divide among cabbage leaves; roll and secure with toothpicks or string. Brown the roll-ups in butter. Pour in water almost to cover. Simmer for about 40 minutes and pour remaining juice over the cabbage rolls to serve.

Variation: Omit the nutmeg and sage, and add ½ teaspoon oregano and 1 cup (205 g) cooked rice to meat mixture. Instead of water, pour 1 can (305 g) tomato soup over rolls and simmer.

Tofu Sea Island Salad

2 cups macaroni or vegeroni, dry (195 g)
½ cup mayonnaise or salad dressing (115 g)
1 cake tofu drained (4 oz/110 g)
½ teaspoon celery seed
½ teaspoon onion salt
1 can tuna or other canned fish, or ham, luncheon meat, chicken
1 can peas, drained (17 oz/476 g)
¾ cup Cheddar cheese, diced (70 g)
1 tablespoon pimiento, diced

Cook macaroni as directed on package. Stir in mayonnaise or salad dressing which is blended with the tofu and seasonings. Gently stir in the peas and remaining ingredients. Chill. Serves 6–8.

Serve with bread or rolls and banana pudding.

Tofu Herb Rolls

2 cakes tofu (½ lb/225 g)
1 package dry yeast
½ cup lukewarm water (118 ml)
2 tablespoons sugar
1 teaspoon salt
½ teaspoon dill weed
1 teaspoon dry onions, minced
pinch of garlic powder
3 to 3½ cups flour (375 to 440 g)
1 egg
2 tablespoons shortening
sesame seeds

Drain tofu. Combine yeast and water in a bowl. Add the sugar, salt, dill weed, onion, garlic powder, and 1 cup (125 g) flour. Mix well. Crumble or mash tofu and stir into the mixture. Stir in egg and shortening. Gradually add remainder of flour until you obtain a dough of kneading consistency. Knead dough until smooth and elastic, about 10 minutes. Place in a greased bowl and cover with a damp cloth. Make sure top of dough is greased too. Let rise until double in bulk, about 2 hours. Punch down and let rest 15 minutes. Then separate into 3 equal balls and roll out into circles about ¼ inch (0.6 cm) thick. Cut into pie shaped pieces and roll up beginning with wide end. Seal narrow edge, forming crescent roll. Roll in sesame seeds. Place on greased baking sheet and let rise again until double in bulk (1 hour). Bake at 375° F (190° C) until browned, about 20 minutes. These rolls remain quite fresh the next day. At least my neighbor thinks so.

Variation: Bake in a loaf bread pan for approximately 45 to 50 minutes or in two 1 lb. (454 g) coffee cans for about 35 minutes at above temperature.

Tofu Coffee Cake

Follow the recipe for making Tofu Herb Rolls but omit the dill weed, onion, garlic, and sesame seeds. When the dough is ready for shaping, spread it out in a 2 × 9 × 13 inch baking pan (5 × 22.5 × 32.5 cm.) which is well greased. Top with Granola (see GRAINS) or sliced fresh peaches dotted with butter and sprinkled with sugar. You may wish to use the Topping recipe given below. Bake at 375° F (190° C) for about 20 minutes or until lightly browned.

Coffee Cake Topping
½ cup sugar (95 g)
¼ cup flour (30 g)
¼ cup butter or margarine (55 g)
 1 teaspoon cinnamon

Mix all ingredients with a pasty blender until they reach a crumbly consistency, then sprinkle on top of dough.

Textured Vegetable Protein (TVP)

GENERAL INFORMATION

Textured vegetable proteins are protein foods which have been developed by extracting the protein from a vegetable source; at present, the most commonly used is the soybean. The proteinaceous material is then forced through tiny holes and dried to give it the texture of meat. One TVP type can be used as an extender and combined with different meats. Others have been developed to replace meat and are called analogs; the analogs have been used by Seventh Day Adventists for a long time.

The extenders are becoming popular in many households. The hamburger-soy mixtures found at meat counters and the dry crumbles available under various names on store shelves are probably the most popular. Textured vegetable proteins can be purchased in various forms, such as canned, frozen, or dried. They can also be purchased unflavored or flavored like beef, ham, etc. The flavored varieties often cost a little more money.

NUTRITIVE VALUE

The textured vegetable proteins are primarily a source of protein and should be treated as such. The quality of the protein is like that of the food from which it is made. The protein quality is below that of meat unless the limiting amino acid is added or it is combined with food that contains enough of the limiting amino acid. No change in the protein quality of a pure meat mixture and a meat mixture containing a soy protein isolate at graded levels of 5 to 25 percent were reported in one scientific study.[1] Therefore, the textured vegetable proteins can make a valuable protein contribution to the diet at less cost than that of meat.

Although the textured vegetable proteins provide a protein source and can be used exclusively in recipes, it is best that they do not replace meat altogether at this time because they do not contain all the micronutrients common to meat. Evidence indicates that the intake of some micronutrients (such as zinc) is marginal in the diets of a growing number of people. These micronutrients, some of which may be unknown, are not included in the enrichment program.

COOKING TIPS

1. Since the textured vegetable protein extenders or analogs look, feel, taste, and chew like meat, fish, ham, or chicken, they can be used in the same manner as these meats or in combination with them.
2. The dried analogs store for a long time and are light in weight. They work well for camping, backpacking, etc.
3. Since the carbohydrate content of a hamburger-soy mixture is greater than that of the ground meat alone, bacterial decomposition, resulting first in flavor changes and then in spoilage, may occur more rapidly. Use butcher-prepared mixtures soon after purchase.

---

[1] Mattil, Karl F. "Composition, Nutritional, and Functional Properties, and Quality Criteria of Soy Protein Concentrates and Soy Protein Isolates," *J. Am. Oil Chemist's Soc. 51:* 81A, January, 1974.

4. To reduce cost and bacterial decomposition, make your own ground meat and soy mixture just before cooking.
5. The following proportion of textured vegetable protein and meat has been approved by the government for use in the school lunch program: 30 percent textured vegetable protein and 70 percent meat. This is probably a good guide to follow.
6. Textured vegetable protein and meat mixtures tend to remain moist during cooking and hold together well. Shrinkage is less since the juice from the meat is absorbed by the TVP. Although the TVP contains little fat and no cholesterol, it does absorb some from the meat.
7. *Special note:* In the recipes given in this section, rehydrated TVP stands for textured vegetable protein granules which have been mixed with water and allowed to stand until the water is absorbed. The amount of water necessary to rehydrate the granules depends on the kind of crumbles being used. Follow any directions given on the package.

Chicken or Turkey Curry

½ cup TVP, unflavored (40 g)
¼ cup water (59 ml)
3 tablespoons margarine
¼ medium onion, minced
1½ teaspoons curry powder
3 tablespoons flour
¾ teaspoon salt
½ teaspoon honey (2.5 ml)
pinch of ground ginger
1 cup milk (235 ml)
1 cup chicken broth (235 ml), or 1 chicken bouillon cube dissolved in 1 cup hot water

1½ cups cooked chicken or turkey, diced (420 g)
½ teaspoon lemon juice (2.5 ml)

Mix TVP with water to rehydrate. Melt margarine over low heat in a heavy saucepan. Sauté the onion and curry. Blend in flour and seasonings. Cook over low heat until mixture is smooth and bubbly. Remove from heat and stir in milk and chicken broth. Bring to a boil, stirring constantly. Cook until thickened and then add chicken, TVP, and lemon juice. Cook until chicken and TVP are heated through. Serves 4.

Spoon over cooked rice. Sprinkle the top with your choice of the following accompaniments: chutney, raisins, tomato wedges, slivered almonds, chopped peanuts, pineapple chunks, bacon bits (real or imitation), flaked coconut, India relish.

Turkey Patties

½ cup TVP, unflavored (40 g)
¼ cup water (59 ml)
¾ pound turkey, ground (340 g)
1 teaspoon seasoned salt
½ teaspoon oregano, ground
3 tablespoons ketchup
2 tablespoons wheat germ
1 tablespoon margarine
2 tablespoons Parmesan cheese, grated

Mix water with TVP to rehydrate. Combine all ingredients except margarine. Shape into 4 patties. Fry in margarine or broil on a barbeque grill. Serves 4.

Serve with three bean salad, buttered noodles, and broccoli, or serve between hamburger buns and garnish as hamburger.

Low Cost Fish Cakes

⅔ cup unflavored TVP (50 g)
⅓ cup water (78 ml)
1 can mackerel (15 oz/420 g)
¼ pound saltine crackers, crushed fine (120 g)
3 large eggs, beaten
1½ teaspoons onion powder or ¼ cup onion, chopped (35 g)
½ teaspoon salt
dash of celery salt
flour
oil for frying

Mix TVP with water. Mash the mackerel. If you don't like the skin and bones, remove them. However, the bones of canned fish are usually soft and are a good source of calcium. Mix in the rehydrated TVP, cracker crumbs, eggs, onion, and salts. Shape into 10 three-inch (7.5 cm) patties, roll in flour, and fry in oil over medium heat until brown. Turn frequently. Serves 5.

Serve on buns with ketchup or tartar sauce, along with a bowl of vegetable soup. You can also serve the fish cakes with lemon slices, tossed green salad, french fries, and a green vegetable.

Calypso Skillet

½ cup TVP, beef flavored or unflavored (40 g)
⅓ cup water (78 ml)
½ pound ground beef (230 g)
½ onion, chopped
1 can corn, drain but save liquid (12 oz/336 g)
1 can tomato sauce (8 oz/227 g)
10 pitted olives, halved
1 teaspoon oregano
½ teaspoon salt
¼ teaspoon pepper
5 ounces uncooked thin noodles (140 g)
1 cup Cheddar cheese, shredded (100 g)

Mix TVP with water to rehydrate. Then mix rehydrated TVP with ground beef. Brown ground beef mixture with onion. Drain off any fat. Add enough water to corn liquid to make 2 cups (474 ml). Stir liquid and remaining ingredients into the meat mixture. Simmer uncovered, stirring occasionally until desired consistency (about 15 to 20 minutes). Serves 4.

Serve with fruit salad, green vegetable, and French bread.

Enchiladas

½ pound ground beef (230 g)
½ cup TVP (40 g)
¼ to ⅓ cup water (59 to 78 ml)
 1 can enchilada sauce, or make your own as given below (19 oz/538 g)
12 corn tortillas
½ medium onion, chopped
 3 oz shredded cheese (80 g)
 1 can olives, pitted and sliced (Dr. Wt. 6 oz/168 g)
 1 cup lettuce, shredded (80 g)
 1 cup yogurt (8 oz/227 g)—optional

Mix TVP with water to rehydrate. When rehydrated combine with ground meat and brown. Heat the enchilada sauce. Dip the tortillas in the enchilada sauce and fill with ground meat mixture, onions, cheese, and olives. Roll up and fasten with toothpick and place in a greased casserole. When all of the enchiladas are complete, pour remainder of the sauce over the top and garnish with shredded cheese. Bake at 350° F (177° C) for about 20 minutes or until cheese is melted. Garnish with shredded lettuce, olive slices, and yogurt.

Variations: Omit the ground meat and use lots of cheese. You may wish to add some cooked pinto beans or kidney beans or soybeans, too. Substitute diced or ground cooked chicken or turkey for the beef and combine with unflavored TVP.

Enchilada Sauce

1 tablespoon flour
1 tablespoon oil (15 ml)
1 teaspoon cumin
1 tablespoon chili powder

1 teaspoon salt
pepper to taste
1 can tomato sauce (8 oz/227 g)
pinch of oregano

Brown flour first; then add chili powder and oil. Add the remainder of the ingredients and bring to boil. Allow to simmer for a few minutes.

This sauce makes enough for 6 enchiladas.

Clam Marinara Sauce

  2 cloves garlic, sliced or smashed
  2 tablespoons oil (30 ml)
  1 onion, diced fine
  2 carrots, diced fine
  1 large can tomatoes (1 lb/454 g)
Italian seasoning, to taste
  1 tablespoon sugar, optional
salt and pepper, to taste
  1 can clams (8 oz/227 g)
½ cup unflavored TVP soaked in the clam juice (40 g)

Sauté the garlic in the oil. Add the onion and carrot and sauté until tender. Add the tomatoes with liquid; crush the tomatoes. Add the Italian seasoning, salt and pepper, and sugar. Cook and simmer for about 1 hour. Just before serving add the clams and the TVP. Pour over spaghetti. Serves 4.

Mexican Casserole

½ pound ground beef, chicken, or turkey (230 g)
½ cup TVP, rehydrated (40 g)
1 medium to large onion, diced
1 medium can refried beans or make your own (20½ oz/574 g)
1 can Cheddar cheese soup (11 oz/312 g)
1 can tomato sauce (8 oz/227 g)
salt and chile powder to taste
10 to 12 corn tortillas

Combine ground beef and TVP and fry with onion. Drain excess grease. Combine beans, cheese soup, tomato sauce, salt, chile powder and meat mixture and heat through. Layer mixture with 10 to 12 corn tortillas in a large baking dish. Bake at 350° F (177° C) for 30 minutes. Serves 6–8.

Tacos

1½ pound ground meat (680 g)
½ cup TVP, rehydrated (40 g)
½ onion, diced
1 teaspoon salt
¼ teaspoon pepper
¼ teaspoon cumin
1 clove garlic, minced
1 small can tomato sauce (8 oz/227 g)
6 to 8 taco shells or pocket bread

Combine ground meat with TVP and fry with onion and garlic. Pour off any excess fat. Combine spices with water and add to meat mixture. Cook for 15 minutes. Add tomato sauce and bring to a boil. Cook an additional 5 minutes. Fill pocket bread or taco shells with the meat mixture. Garnish with tomatoes, grated cheese, and alfalfa sprouts or lettuce. Serves 6–8.

Meat and Potato Roll-Up

  2 potatoes, cooked and mashed
  1 tablespoon parsley, snipped
  ⅛ teaspoon thyme
  ⅛ teaspoon marjoram
  2 tablespoons Cheddar cheese, grated
  1 pound lean ground meat (454 g)
  ½ cup TVP, rehydrated (40 g)
  1 egg, beaten
  ⅓ cup tomato sauce (78 ml)
  ¼ cup onion, chopped fine (35 g)
  ¼ cup green pepper, chopped fine (35 g)
salt to taste

First cook and mash the potatoes so that they are ready. Stir the parsley, salt, thyme, marjoram, and cheese into the potatoes. Set aside. Mix together the meat, TVP, egg, tomato sauce, onion, green pepper, and salt. Flatten into a thick rectangle. Place the potato mixture across the narrowest part. Wrap the meat around the potatoes, sealing the edges. Place on a baking pan or rack. Bake at 350° F (177° C) for 45 minutes or until done. Slice to serve. Serves 6.

Disaster to Delicious

A student prepared her own modified version of the Meat and Potato Roll-up by substituting a tablespoon of dried onion and a tablespoon of dried green pepper and two cups of prepared instant potatoes. She made two large meat patties, the size of a pie-pan, placing the potatoes between but failed to seal the edges. Thirty minutes later she checked the recipe and found the potatoes and juice running out. To save the recipe she removed the meat patties to a clean pie-pan, placing the potatoes on top of the patties and sprinkling with paprika. She placed this under the broiler long enough to dry out the potatoes. She cut it into pie shaped pieces and served with cauliflower, zucchini, and a green salad. She said it was the best meat loaf she had ever made.

Ground Beef Pie

Crust

 1 cup biscuit mix (140 g)
 ¼ cup margarine (55 g)
 3 tablespoons hot water (45 ml)

Combine all ingredients and pat into a 9-inch (22.5 cm) pie pan.

Filling

 ¾ lb. ground meat (350 g)
 ½ cup TVP, rehydrated (40 g)
 ½ small onion, chopped
 ½ teaspoon salt
 ¼ teaspoon pepper
 1 tablespoon Worcestershire sauce (15 ml)
 2 tablespoons biscuit mix
 2 eggs
 2 tablespoons milk (30 ml)
 ½ cup Cheddar cheese, shredded (45 g)

Mix together the ground meat and rehydrated TVP. Brown the meat mixture with the onion. Add salt, pepper, Worcestershire sauce, and biscuit mix. Stir. Fill the pie crust. Beat eggs and milk slightly. Add the cheese and pour over the meat mixture. Cover the crust edge with strips of foil. Bake at 375° F (190° C) for 30 to 40 minutes. Serves 4–6.

Serve with a green vegetable, fruit, and cookies for dessert.

Tournament Supreme

A friend of mine used to serve this to a crowd during basketball tournament time in Indiana.

  1 onion, chopped
  ¾ lb. ground meat (350 g)
  ½ cup TVP, rehydrate after measuring (40 g)
  ½ green pepper, chopped
  3 stalks celery, chopped
  1 can tomato soup (10¼ oz/305 g)
 1¼ cups water (295 ml)
  1 can mushrooms (2 oz/56g)

Brown the ground meat and rehydrated TVP with the onion. Add green pepper, celery, tomato soup, and water and simmer for 45 minutes. Add the canned mushrooms and simmer about 10 more minutes. Serves 8.

Serve over Chinese noodles.

Cannelloni

Filling

  ¼ lb. ground meat (115 g)
  ½ cup TVP, measure and then rehydrate (40 g)
  1 medium onion, diced fine
  ½ teaspoon garlic powder
  6 large mushrooms, chopped fine
  2 tablespoons Parmesan cheese, grated
  1 bunch spinach or swiss chard, chopped and cooked
      or 1 package (10 oz/280 g) frozen chopped spinach
  1 cup bread crumbs (50 g)
  3 eggs, slightly beaten
 1½ teaspoons salt
  ¼ teaspoon pepper
 16 cannelloni pastas

Brown the beef mixed with the TVP, onion, garlic, and mushrooms in oil for about 5 minutes. Mix with other ingredients. Fill the uncooked cannelloni pastas with filling.

Sauce

- 1 medium carrot, grated fine
- 1 tablespoon parsley, snipped
- 2 tablespoons onion, minced
- 5 small cans of tomato sauce (8 oz/227 g each)
- ½ teaspoon pepper
- ½ teaspoon basil

Sauté carrot, parsley, and onion for about 10 minutes. Add tomato sauce, pepper, and basil and simmer for about 1 hour. Add water if you need to. When sauce is finished, cover bottom of oblong casserole with sauce. Place the cannellonis in the sauce. Cover cannellonis with remaining sauce and add ½ cup (118 ml) water. Cover casserole with aluminum foil and bake at 375° F (190° C) for about 1 hour. Serves 8.

Legumes

GENERAL INFORMATION

The dictionary defines legumes as the fruit or seed of a pod-bearing plant, such as beans, peas, and lentils, used for food. The soybean is a legume but since its protein value is higher than that of other legumes, the preceding chapters were devoted to it and two of its products, tofu and textured vegetable proteins. Peanuts are a legume, but they are included in the chapter on Grains, Nuts, and Seeds.

NUTRITIVE VALUE

In general, legumes contain a larger percentage of protein than other plant foods. A ⅔ cup (120 g) serving of most of the mature, dried beans and peas yields about 9 to 10 grams of protein after cooking. This is roughly ⅕ of a 50 gram daily requirement. The dried beans and peas are rather high in starch carbohydrate but are very low in fat and contain no cholesterol. They do not contain any vitamin A, vitamin C (ascorbic acid), or vitamin B_{12}. However, they do provide useful amounts of the other B vitamins and minerals such as iron. The iron contribution can be increased if they are cooked in an old iron pot.

Many beans and peas can be grown in home gardens. The fresh pea or bean appears to have a lower water content after cooking, and the concentration of nutrients is therefore a little greater on a given weight basis. A small amount of vitamin A value and ascorbic acid is present in fresh peas and beans.

COOKING TIPS

1. Soak dried beans and peas to cut down the cooking time. Two methods are recommended. One is to wash the beans, boil them for 2 minutes, remove from heat, soak 1 hour, and then cook. The other is to let them soak overnight in cold, salted water to cover.
2. To cook, either drain the beans and add fresh water or cook in the soaking water. Beans should be well covered with water and allowed to boil gently for about 2 hours. Add 1 teaspoon salt for each cup (180 g) of dried beans or peas. Long slow cooking helps to reduce the formation of gas in the intestinal tract. A student told me that she sprouts all legumes before cooking and no longer has gas problems.
3. If your water is extremely hard, add ⅛ teaspoon (0.5 g) soda for each cup (180 g) of dry beans. Do not add more than this or the B vitamins, especially thiamin, will be destroyed.
4. The addition of fat such as meat drippings, salt pork, ham hock, or oil reduces foaming during cooking and adds flavor.
5. Use a pressure cooker to reduce cooking time to less than 30 minutes after soaking. Check your pressure cooker's instruction booklet for complete directions.
6. Lentils may or may not be soaked before cooking. They do not require as long a cooking time as other legumes.

7. Soaking split peas for ½ hour after a 2 minute boil will help to retain their shape. It is best not to cook them in a pressure cooker because they may splatter and clog the vent.
8. Ingredients such as tomatoes, ketchup, or vinegar should be added after the vegetables are tender, since the acid will prevent the vegetables from softening.
9. Cook fresh beans and peas about 20 minutes or until tender.
10. For use in salads, cook beans only until tender and drain immediately because their liquid thickens fast.
11. To mash beans, cook the beans until very tender and mash or puree while hot.
12. Pureed beans and bean dishes freeze very well so cook a lot at one time to conserve energy.

Basic Lentils

This is the basic method for preparing lentils before use in other recipes.

1½ cups lentils (270 g)
  2 cups water (470 ml)
  1 cup broth prepared from ham bone (235 ml)
  1 onion, minced
  2 bay leaves

Wash the lentils well, add the water and soak for 6 hours or overnight. When ready to cook, add broth, onion, bay leaves, and salt. Simmer until tender. Add a little boiling water if they become too dry before they are done. Lentils are now ready for measuring and using in a loaf, pureed or other lentil dish.

Stewed Lentils

1½ teaspoons salt
¼ onion, minced and sautéed
¼ green pepper, minced
3 tablespoons oil (45 ml)
1 tablespoon lemon juice (15 ml)

Prepare Basic Lentils and do not drain. Add the salt, green pepper, sautéed onion, and lemon juice.

Lentil Puree

1½ cups lentils prepared according to Basic Recipe (270 g dry)
1 carrot, shredded
1 stalk celery, chopped
1 teaspoon salt
1 onion, minced
¼ teaspoon pepper
2 tablespoons lemon juice (30 ml)
2 tablespoons butter or margarine
¼ cup cream or tomato puree (59 ml)
¼ cup whole wheat bread crumbs (25 g)
¼ cup cheese, grated (23 g)

Prepare lentils according to basic recipe and cook the carrots, celery, salt, onion, and pepper together. Mash the lentils and vegetables through a sieve. Add the lemon juice, butter, and cream or tomato puree. A few drops of Worcestershire sauce and a teaspoon of syrup may help the flavor. Put in a greased baking dish and cover with the crumbs and cheese. Heat and brown in the oven at 350° F (177° C) for about 20 minutes.

Green Peppers Stuffed with Lentils

Green peppers
Stewed lentils or lentil puree
Pork sausage

Select the required number of peppers of uniform size, wash them, cut off the tops, and remove the seeds and white pith without breaking them. Parboil for 2–3 minutes, drain. Make a lentil puree or follow the recipe for stewed lentils. Have as many little pork sausages as peppers. Brown them a little. Fill each pepper with one sausage and the lentils, packing them around the sausage. Bake 25 minutes at 350° F (177° C). Put a little water in the bottom of the pan to prevent peppers from burning.

Lentil Loaf with Tomato-Cheese Sauce

1½ cups dried lentils prepared according to basic recipe
    (270 g dry)
salt and pepper to taste
  1 onion, minced
  1 green pepper, minced
  1 tablespoon olive oil (15 ml)
  2 teaspoons oregano
  2 tablespoons lemon juice (30 ml)
  1 cup soft bread crumbs (100 g)

Wash the lentils and soak them in cold water overnight; drain. Prepare lentils according to basic recipe, add salt and pepper to taste. Sauté the minced onion and green pepper in olive oil, and add to the lentils with the other ingredients. Put into a greased, floured bread pan and bake 45 minutes at 375° F (190° C). Serve on a platter with the tomato-cheese sauce poured on top.

Tomato-Cheese Sauce

  1 onion, minced fine

½ green pepper, minced fine
3 tablespoons olive oil (45 ml)
1⅓ cups tomato puree, canned (313 ml)
½ cup sharp cheese, grated (46 g)
salt and pepper
1 teaspoon syrup (5 ml)
1 teaspoon fennel (optional)

Fry the finely minced onion and pepper in the oil for 4–5 minutes, then add the tomato puree and simmer for 10 minutes. Add the other ingredients and stir until the cheese is melted but do not cook any more. This is a good sauce to use with lentil dishes.

Spiced Baked Lentils

1 package lentils (1 lb/454 g)
1 teaspoon salt
3 cups water (705 ml)
4 slices bacon, diced
2 cans tomato sauce (454 g)
1 medium onion, chopped
¼ cup brown sugar (52 g)
2 tablespoons prepared mustard
⅓ cup molasses (78 ml)

Sort through lentils, rinse, and drain. Place lentils in a covered casserole (2½–3 quart). Add salt, water, diced bacon (uncooked), tomato sauce, onion, brown sugar, mustard, and molasses. Stir well. Cover dish; bake in a 350° F (177° C) oven until lentils are soft and the liquid is thick and bubbly, about 2 hours. Stir about every 30 minutes while baking, adding a little more water if sauce becomes too thick. One pound (454 g) of turkey hot dogs cut in 1-inch (2.5 cm.) slices might be added during last 30 minutes for a hearty one dish meal. Serves 8.

Lentil Vegeburger

¼ cup pearl barley (45 g)
¼ cup lentils (45 g)
2 stalks celery, diced fine
2 carrots, diced fine
1 small onion, diced fine
6 large button mushrooms, sliced
½ bunch parsley, chopped
4 tablespoons melted butter (60 ml)
3 eggs
salt and pepper to taste
4 slices Swiss cheese

Combine barley and lentils and boil in water for 40 minutes. Dice celery, carrots, and onion. Slice mushrooms and chop parsley. Pour melted butter into hot frying pan. Immediately add celery, carrots, onion, and mushrooms and sauté in butter (should take 3 to 5 minutes).

In mixing bowl, combine sauteed mixture with boiled barley and lentil mixture and stir. Add the eggs, parsley, salt and pepper to taste. Stir until evenly mixed. Pour mixture into round or oval baking tins, the size of desired burger. Place tins in oven (pre-heated to 350° F/177° C) and bake for 15 minutes. Remove from oven, top each burger with a slice of Swiss cheese and bake for 7 minutes more. Serves 4.

Spareribs with Lentils

Spareribs

6 to 7 pounds spareribs (2,724 to 3,178 g or 3 kg)
2 to 3 tablespoons soy sauce (30 ml to 45 ml)
1 tablespoon salt
2 teaspoons cracked black pepper

Rub the spareribs with soy sauce, salt, and pepper on both sides. Place on a rack and bake at 350° F (177° C) for approximately 1 hour, turning once, or until spareribs are brown and crispy. Cut into sections and serve on a hot platter with sauteed lentils and relishes—mustard, horseradish, chow-chow.

To Cook Lentils

1 lb. lentils (2½ cups/454 g)
1 tablespoon salt
1 bay leaf
1 onion stuck with 2 cloves
2 or 3 garlic cloves
7 cups water (1,645 ml or 1.6 l)

Place lentils, salt, and other seasonings in large saucepan and cover with the 7 cups (1.6 l) water. Bring to boil, and boil 2 minutes. Let stand for 1 hour. Bring to a boil again and simmer about 15 minutes, or until lentils are soft, not mushy. Drain well. Remove bay leaf and onion.

Instead of recipe above, you can follow directions on package.

To Sauté Lentils

6 to 8 slices bacon
4 tablespoons bacon fat (60 g)*
3 to 6 tablespoons butter or margarine (45 g to 70 g)*
2 medium onions, chopped
2 cloves garlic, minced
½ cup parsley, finely chopped (10 g)
black pepper to taste

Sauté the bacon until crisp. Drain on paper toweling; break into tiny pieces. Remove all but 4 tablespoons of the fat from the skillet; add butter or margarine. Heat the fat and sauté the onion and garlic until just golden. Add the lentils, bacon bits and chopped parsley. Toss gently with a spoon and fork and serve. Makes 6 servings.

---

* Total amount of fat to be used about 6 tablespoons (70 g). Use a combination of bacon fat and margarine, or 6 tablespoons of margarine.

Falafel

1 can garbanzos, ground (15 oz/425 g)
1 can fava beans (lima or soy beans may also be used), ground (454 g)
1 medium onion, diced
4 tablespoons flour (30 g)
4 garlic cloves, minced
2 teaspoons salt
¼ teaspoon black pepper
2 teaspoons cumin, ground
½ teaspoon coriander (optional)
1 teaspoon baking powder
1 bundle parsley
sesame seed sauce
pocket bread (page 111)

Mix the ingredients together and cook like meatballs. Put in pocket bread and garnish with tomatoes, lettuce, olives, and sesame seed sauce. Hot sauce is optional.

Sesame Seed Sauce

Purchase Sesame Tahini or make your own sesame paste by spreading 1⅓ cups (250 g) sesame seeds in shallow pan and toasting at 350° F (177° C) until lightly browned (10–15 minutes). Then blend at high speed until powdery. Gradually add ⅓ cup (78 ml) oil and blend until smooth. Add ¼ teaspoon salt. Makes 1 cup (270 g).

1 cup sesame seed paste or Sesame Tahini (270 g)
2 cloves garlic, crushed
1 teaspoon salt
dash cumin
¼ cup lemon juice (60 ml)
¾ to 1 cup cold water (175 ml to 235 ml)

Mix sesame seed paste, garlic, seasonings, and lemon

LEGUMES

juice. Beat in ½ cup (118 ml) of water; beat in remainder, a tablespoon at a time, until sauce has the consistency of mayonnaise. Makes about 1½ cups (350 ml). Store in refrigerator until used.

Dried Beans and Dried Peas (Any Kind)

Soak the beans overnight. Then cook several hours with a ham hock and salt. Serve with bread of some kind.

You can also bring the beans or peas to a boil and let them sit for an hour; then boil them gently for several hours.

You can also soak the beans overnight; drain them and wash them about 3 times a day until they begin to sprout; then cook them.

Fava Beans with Tomato Sauce

2 cups fava beans, shelled (460 g)
1 small can tomato sauce (227 g)
1 clove garlic, minced
1 small onion, chopped
½ green pepper, chopped
salt to taste
2 tablespoons oil, margarine, or bacon drippings (30 ml)

Cover fava beans with water and add salt. Cook until just tender. Sauté the onion, garlic, and green pepper in the fat. Add the tomato sauce and simmer a few minutes. Add 2 tablespoons of the cooked fava beans to the mixture and simmer about 5 minutes. Add the simmered mixture to the remainder of the drained fava beans and serve with a hot bread. Serves 4.

The fava bean, also called a horsebean, is a favorite of many Portuguese people. It resembles a lima bean only it is larger. There is a skin around the bean itself which is removed after cooking and before eating.

Pinto Beans—Portuguese Style

1 small onion, chopped
½ green pepper, chopped
2 tablespoons oil or margarine (30 ml)
1 teaspoon cinnamon
2 tablespoons brown sugar
1 can tomato sauce (227 g)
2 cups pinto beans, cooked (460 g)
salt to taste

Sauté the onion and green pepper in the oil until transparent. Combine the green pepper, onion, cinnamon, brown sugar, and tomato sauce. Bring to a boil and add 2 spoonfuls of beans. Simmer for 3 minutes; then add the remainder of the beans and heat through. Serve with a hot bread. Serves 4.

Pinto Beans

Soak a one pound (454 g) package Pinto Beans by one of the methods mentioned under Dried Beans and Dried Peas.

salt to taste
1 can tomato sauce (227 g)
1 tablespoon chili powder
½ onion, diced
ham hock (optional)

Refried Beans

2 tablespoons bacon grease (30 ml)
2 cups Pinto beans, cooked and mashed (460 g)
salt and pepper to taste
dash of Tabasco sauce or diced green chili
½ cup Monterey Jack cheese or Cheddar cheese (46 g)

Spread a casserole dish with bacon grease. Add the mashed beans mixed with seasonings. Top with cheese and bake for 20 minutes at 350° F (177° C). Serves 4.

Serve with tortillas or in sopaipillas (page 110).

Three Bean Casserole

  1 can pork and beans (1 lb/454 g)
  1 package frozen lima beans (10 oz/280 g)
  1 can kidney beans (15¼ oz/432 g)
  1 onion, chopped
 ¼ cup ketchup (59 ml)
 ¼ cup brown sugar (52 g)
1½ teaspoons prepared mustard
  1 teaspoon soy sauce (5 ml)

Combine all ingredients in skillet or casserole and cook until most of liquid is evaporated (several hours). Add salt and pepper to taste. Meat may be added. Serve with cornbread. Serves 4–6.

Three Bean Casserole with Cheese

1 onion, chopped
1 can baked beans (1 lb/454 g)
1 can kidney beans (15¼ oz/432 g)
1 package frozen lima beans (10 oz/280 g)
½ cup brown sugar (53 g)
⅓ cup ketchup (78 g)
½ teaspoon Worcestershire sauce (2.5 ml)
1 cup cheese (your choice) cubed or grated (92 g)

Mix all ingredients except cheese, which is put on top. Bake at 325° F (163° C) for 30 minutes. Serves 4–6.

Chicken and Garbanzo Beans (Chick Peas)

- 1 frying chicken cut up, or enough breasts, thighs, or wings to serve 6
- 6 tablespoons margarine (90 ml or 70 g)
- ½ onion, chopped fine
- 1½ teaspoons seasoned salt
- ¼ teaspoon ginger, ground
- ¼ teaspoon cumin, ground
- ¾ teaspoon black pepper, ground
- 1 can garbanzo beans (15 oz/425 g)
- 2 chicken bouillon cubes dissolved in 1¾ cups hot water (415 ml)

Remove skin and then brown chicken in margarine. Remove chicken and add onions and sauté until limp. Return chicken to skillet and add salt, ginger, cumin, pepper, beans, and bouillon. Cover and simmer about 45 minutes or until liquid has cooked down and chicken is tender. Serves 6.

Black-Eyed Peas and Rice

- 6 cups water (1410 ml or 1.4 l)
- 2 cups dried black-eyed peas (320 g)
- 1 onion, chopped
- ½ lb. bacon, diced or ½ lb. salt pork (230 g)
- ¼ teaspoon cayenne pepper or 1 dried hot red pepper (optional)
- 1 teaspoon salt or to taste
- 1 cup white or brown rice, uncooked (180 g)

Wash the black-eyed peas and soak by one of the methods given on page 69. Add onion, bacon, and red pepper. Bring to a boil and simmer until peas are tender, about 1 hour. Add salt and rice and more liquid if necessary.

Cook until rice is done (about 15 minutes for white and 30 minutes for brown). Do not overcook. Serves 4–6.

Serve with small boiled turnips cooked with the greens and cornbread for a real Southern meal. The rice complements the protein in the black-eyed peas and improves the protein quality.

Variation: The black-eyed peas are also good served with fried okra. Slice the okra, sprinkle with salt and place in a paper bag containing cornmeal. Shake the okra in the bag. Remove from bag and fry in bacon drippings or oil until browned.

Grains, Nuts, and Seeds

GENERAL INFORMATION

A dictionary definition of grains is the seed or seedlike fruit of any cereal grass such as wheat, maize, corn, oats, or rice. In many parts of the world these grains, especially rice, wheat, and corn, provide a major source of food and amino acids. They are relatively cheap, easy to grow, easy to store, and can add variety as well as essential nutrients to our diets when consumed in sensible amounts.

Grains can be obtained in a variety of forms, ranging from the whole grain to the refined form, and all varieties can have a function in our diets.

NUTRITIVE VALUE

Grains provide a large amount of carbohydrate in the form of starch. This carbohydrate is an important and inexpensive source of energy and provides a less favorable environment for dental cavities than does sugar, another carbohydrate. We need some carbohydrate for energy (instead of using protein) and to prevent the too-rapid burning of fat which leads to ketosis, a state which exists in uncontrolled diabetes. In addition, the cereals or grains provide valuable amounts of B vitamins, minerals (especially iron), and a small amount of protein.

Table VII
Nutrient Composition of Selected Grains, Nuts, and Seeds, 100g [1]

Common Measure	Food	Calories	Protein g	Carbo-hydrate g	Fat g	Calcium mg	Iron mg	Vita-min A I.U.	Thia-min mg	Ribo-flavin mg	Nia-cin mg	Vitamin C mg
3/4 c.	Whole wheat, dry	340	9.9	76.2	2.0	36	3.2	0	.36	.12	4.1	0
1 c.	Oats, dry	390	14.2	68.2	7.4	53	4.5	0	.60	.14	1.0	0
3/5 c.	Corn	96	3.5	22.1	1.0	3	.7	400	.15	.12	1.7	12
1/2 c.	Rice, dry brown	360	7.5	77.4	1.9	32	1.6	0	.34	.05	4.7	0
2/3 c.	Peanuts	564	26.0	18.8	49.8	74	2.1	—	1.14	.13	17.2	0
2/3 c.	Sesame seeds	582	18.2	17.6	53.4	110	2.4	—	.18	.13	5.4	0
3/5 c.	Sunflower seeds	560	24.0	19.9	47.3	120	7.1	50	1.96	.23	5.4	—
1 c.	Walnuts, English	651	14.8	15.8	64.0	99	3.1	30	.33	.13	.9	2

[1] Values taken from Composition of Foods, Agriculture Handbook No. 8, Agricultural Research Service, United States Department of Agriculture, 1963.

The whole grains, such as brown rice and whole wheat, contain more fiber and fat than the processed form which has had the bran and germ removed. The grains are a good source of vitamin E as long as the germ is present and has not been removed. In general, the grains, except for fresh corn, lack vitamin A and vitamin C.

The nuts and seeds contain a lot more fat and a lot less carbohydrate than the grains. The protein concentration is also a little greater. However, the amino acid weakness of the nuts and seeds is very much like that of the grains so I am including them in this section.

In general, the grains [1] nuts, and seeds tend to be deficient in lysine in particular, but rather strong in methionine. As a result, it is best to combine them with milk products, fish and meats, or legumes.

COOKING TIPS

1. Whole wheat is good either steamed or cooked; however, the cooking process takes a long time. Bulgur, which is parched and cracked whole wheat, has been a staple in the Middle East for many years. It requires a rather short cooking time, is delicious, and offers everything that whole wheat offers. A method for making your own is included in the recipe section. Look for bulgur in the rice section or cereal section of your store.
2. There are a variety of rices which can be purchased: long grain, short grain, pearl, wild rice, etc. You will have to determine which you prefer to use. Rice can be purchased as brown, which has the bran and germ remaining; or as converted, which has been steamed to force many of the nutrients from the bran layer into the endosperm before removal of the bran. Regular white rice has the bran removed and should be enriched. Wild rice is good

---

[1] Corn is also quite deficient in tryptophan.

GRAINS, NUTS, AND SEEDS

but quite expensive and does not really belong in a low-cost book unless only small amounts are used in combination with other rice.

3. Gluten is the protein found in the endosperm. Hard, red, winter wheat provides the most. Although it is somewhat deficient in lysine, recipes made with it provide a good protein source when combined with some food high in lysine. Simulated chicken steaks, hamburger, etc. can be made from gluten. Making gluten is a laborious process but directions are given for those ambitious cooks who would like to make it. Canned gluten can sometimes be purchased in specialty sections in stores.
4. Purchase whole wheat spaghetti and macaroni for variety. These products are more expensive than the regular variety, and they do not absorb as much water or get as tender.
5. It is difficult to store whole grain products since the fat present in the germ becomes rancid on exposure to air and warm temperatures. They are also susceptible to insect infestation.
6. Whole wheat kernels can be bought in specialty food shops or health food stores.

Steamed Wheat

2 cups wheat (400 g)
1¾ cups water (412 ml)
1 teaspoon salt

Put ingredients in a small pan or casserole dish (the wheat will increase 2½ times in size). Set it on a rack in a larger pan. A tuna can with holes punched in it will be suitable for a rack. Pour water into the large pan to the level of an inch (2.5 cm) and then place the small pan in the large pan. Cover and steam 4 hours. Reduce heat after first 15 minutes.

Boston Baked Wheat

3 slices bacon, diced in large pieces
1 onion, diced
4 cups steamed wheat (440 g)
1 cup ketchup (235 ml)
1 cup water (235 ml)
½ cup mild molasses (118 ml)
3 to 4 teaspoons prepared mustard
½ cup soy grits (70 g), soaked for 1 hour

Fry bacon and sauté the onion in the grease. Then mix all the ingredients together and add salt and pepper to taste. Bake at 325° F (163°) for ½ hour. Stir occasionally.

Preparation of Bulgur (Cracked Wheat)

4 cups whole wheat (800 g)
water to cover

Wash the wheat in about 4 changes of cool water. Then put the wheat in a pot with enough water to cover. Boil until all the water is absorbed and the kernels are tender and double in size, about 35 to 45 minutes. Then spread the kernels on a cookie sheet and dry it completely in a 200° F (93° C) oven. Put the wheat through a grinder or mill to crack it into smaller pieces the size of regular cracked wheat. Store it in an air tight container in a cool place, preferably the refrigerator.

Bulgur has quite a variety of uses. It can be cooked for 5 to 10 minutes and eaten as a breakfast cereal. Use it in breads, rolls, cookies, salads, or in hamburger to extend it.

Bulgur Pilaff

2 tablespoons oil (30 ml)

1 green onion, chopped with leaves
½ onion, chopped fine
1 cup bulgur (165 g)
2 cups water (470 ml)
2 chicken or beef bouillon cubes
salt and pepper to taste

Sauté the onions in the oil until transparent. Add the bulgur and lightly brown it.* Then add water, bouillon, and seasonings. Simmer 15 to 20 minutes until all liquid is absorbed. Serve with chicken or meat kabobs. This pilaff is also good served as an accompaniment to eggs at breakfast or any other meal. Serves 4 to 6. This is one of my family's favorite side dishes.

Variation: Add browned hamburger and grated cheese and bake until cheese melts.

Preparation of Gluten

10 cups whole wheat flour (1200 g or 1.2 kg)
6 cups water (1410 ml or 1.4 l)

Mix together the flour and water. The dough should run through your fingers and shouldn't be too stiff or too loose. Knead in a heavy duty mixer for 10 minutes. (A little stiffer dough can be kneaded by hand but it takes a long time to develop the gluten.) After kneading is completed, place dough in warm water and knead for 3 minutes by hand. It's best to do this in the sink. Then place the gluten in a colander and pour water over it, about a cup at a time, while continuing to knead. After all the starch is washed out, put the gluten into 2 greased and floured loaf pans. Bake at 350° F (177° C) for 1 and ½ hours. It will keep in the refrigerator and freezes well.

Before use in a recipe, gluten must be seasoned, simmered at least ½ hour and soaked for 2 hours or overnight.

---

* Browning too long at high temperatures destroys the thiamin.

Gluten Steaks—Chicken Fried

1 gluten loaf, baked
4 cups water (940 ml)
4 beef or chicken bouillon cubes
2 eggs, beaten
1 cup cracker crumbs (59 g)

Slice the gluten loaf and simmer for ½ hour in the water containing the bouillon. Soak for 2 hours or overnight. Drain in colander to remove excess juice. Then dip the gluten slices in the eggs and the cracker crumbs and fry in a small amount of oil. Serves 8.

Gluten Won Tons

1 gluten loaf, baked
4 cups water (940 ml)
4 chicken bouillon cubes
2 green onions, chopped fine
1 sprig parsley, chopped fine
6 canned water chestnuts, chopped
1 tablespoon soy sauce (15 ml)
¼ teaspoon ginger, ground
pinch of poultry seasoning, optional
1 package won ton skins

Simmer the gluten loaf in the bouillon and water for ½ hour and allow to soak for at least 2 hours. Then chop it fine to resemble hamburger. Drain off excess juice in colander. Combine all ingredients except won ton skins. Place 1 teaspoon of chopped gluten mixture in the center of a won ton skin. Roll as in egg roll, sealing edges with water. Fry in deep fat until golden brown. Serve with Sweet and Sour Sauce (page 46) or Hot Mustard Sauce.

GRAINS, NUTS, AND SEEDS

Hot Mustard Sauce

¾ cup chili sauce (177 ml)
1 tablespoon dry mustard
2 tablespoons water (30 ml)

Blend all ingredients. Use for a dip for won tons and seafood.

Granola

3 tablespoons oil (45 ml)
½ cup brown sugar (53 g)
¼ teaspoon salt
¼ cup water (59 ml)
3 cups regular rolled oats (300 g)
2 tablespoons soy grits or soy flour
½ cup sunflower seeds or nuts (80 g)
¼ cup wheat germ (20 g)

Mix oil, sugar, salt, and water. Heat until sugar dissolves. Cool slightly. Combine oats, grits, and sunflower seeds and place in shallow baking pan. Pour the oil mixture over the cereal mixture and toss gently to blend. Spread to about a ¼ in. (0.6 cm) layer. Bake at 250° F (121° C) for 25 to 30 minutes until mixture is golden brown. Stir once or twice during baking period. After baking, stir in the wheat germ. When cool, store in a covered container in a cool place.

Variation: Add chopped dates, chopped apricots, raisins, etc.

Granola Cookies

⅔ cup brown sugar (35 g)
½ cup butter or margarine (114 g)
2 tablespoons milk (30 ml)
2 eggs, beaten
1½ teaspoons vanilla (7.5 ml)
¼ teaspoon soda
1 teaspoon baking powder
½ teaspoon salt
1 cup flour (123 g)
3 cups granola (360 g)

Cream the sugar, margarine, milk, and eggs thoroughly. Then add the vanilla, soda, baking powder, salt, and flour. Mix thoroughly and stir in the granola. Drop by spoonfuls on greased cookie sheet. Bake at 350° F (177° C) for 8 to 10 minutes. Makes 36 to 48 cookies.

Brown Rice and Beef

3 strips bacon, diced
2 tablespoons onion, chopped
½ pound ground meat (227 g)
1½ cups brown rice (300 g)
1 can consomme (301 g)
1 cup hot water (235 ml)

Fry bacon and drain, leaving about one tablespoon drippings. Brown onion in remaining bacon grease. Add ground meat and brown. Add rice and brown. Add consomme and water. Cook 1 hour. If it becomes too dry, you may have to add more hot water. Serves 4.

Rice Casserole

1 package white and wild rice or ¾ cup brown rice (160 g)

1 small zucchini, chopped
2 stalks celery, chopped
9 medium mushrooms, washed and sliced
1 small onion, chopped fine
½ bay leaf
¼ cup raw sunflower seeds (40 g)
⅛ cup slivered almonds (30 g)
1 egg
1 cup Cheddar and/or Monterey Jack cheese, grated or shredded, enough to top casserole (92 g)

Cook the rice according to directions on the package. Sauté zucchini, celery, onions, garlic, mushrooms and bay leaf in oil or margarine until tender. Combine rice and sautéed ingredients. Add sunflower seeds, almonds and egg. Stir together and pour into a greased casserole. Top with cheese. Bake at 325° F (163° C) for 30 minutes or until heated through and the cheese is melted. Serves 4.

Tabouli

1 cup bulgur (165 g) + 2 tablespoons soy grits (30 ml)
1 bunch parsley, chopped fine
1 green onion, chopped fine
1 small green pepper, chopped fine
1 cucumber, chopped fine
3 tomatoes, chopped fine
2 lemons, juiced
½ cup oil (118 ml)

Soak the bulgur and the soy grits for 1 hour. Drain well and squeeze with hands. Add the chopped vegetables to the wheat. Add the lemon juice, oil, and salt and pepper to taste. Mix well. Serves 4.

This salad is better after standing several hours. Also you may add a little Italian dressing if you wish.

Chipped Beef Casserole

   2 cans condensed cream of mushroom soup (601 g)
   2 cups whole milk (470 ml)
   2 cups Cheddar cheese, grated (184 g)
  ¼ cup onion, chopped fine (40 g)
   2 cups uncooked elbow macaroni (240 g)
2¼ packages dried chipped beef, cut in bite size pieces
   4 hard cooked eggs, diced

Mix soup to creamy consistency, add milk, cheese, onion, uncooked macaroni, and chipped beef. Fold in eggs. Turn into buttered three quart (3 l) baking dish. Store in refrigerator 3 to 4 hours (or overnight). Heat oven to 375° F (190° C) and bake one hour uncovered. It may take two hours because it is so cold. Serves 8–10.

Macaroni and Bacon

   1 package macaroni, cooked (454 g)
  ½ package bacon (227 g)
   1 onion, diced
   1 can tomatoes (454 g)
salt and pepper to taste, and any other spices desired

While the macaroni is cooking, fry the bacon until crisp. Then remove the bacon and sauté the onion. Pour the tomatoes into the onions. Add spices, bacon, macaroni and let set for several hours to absorb flavor. Then reheat to serve. Serves 4–6.

Spanish Style Rice Pilaff

1 small onion, cut fine
2 tablespoons oil (30 ml)
1 cup rice (180 g)

½ cup tomato sauce (118 ml)
1 teaspoon salt
2 cups chicken or beef broth, or water with bouillon cube (470 ml)

Sauté the onion in the hot oil; add the rice; stir constantly until rice starts to brown; add tomato sauce, salt and liquid. Bring to a boil and then turn to simmer and cover. If desired rather dry, the lid should be taken off for the last few minutes of cooking. Cook 15–20 minutes. Serves 4.

Variations: Add chopped bell pepper or green chiles and hamburger, tuna, etc. Top with grated cheese.

Fettucine with Cream

This is also called Fettuccine Alfredo and was served as early as 1200 A.D. in Rome.

1 lb. fettuccine (454 g)
1 cube butter or margarine
½ cup Parmesan cheese, grated (46 g)
milled black pepper or cracked black pepper
2 egg yolks
½ cup heavy cream, warmed (118 ml)
salt to taste

Have all ingredients ready and at hand. Cook the fettuccine and drain. Melt butter in chafing dish or electric pan; add cooked fettucine; toss gently. Grate in cheese; toss, grate in pepper; toss. When cheese is well mixed in, add eggs and toss again, then add warmed cream and toss. Serve immediately. Serves 4–6.

Tomato Macaroni Casserole

    4 slices bacon
    1–8 oz package elbow macaroni (227 g)
    3 tablespoons butter
    3 tablespoons flour
1½ teaspoons salt
pinch of pepper
    2 cups milk (470 ml)
2¾ cups sharp Cheddar cheese, grated (268 g)
    2 medium tomatoes, sliced
paprika

Cook bacon until crisp. Drain and cool. Cook macaroni in boiling salted water until tender (8–10 minutes). Drain. Melt butter in saucepan, add flour, salt, and pepper. Add milk gradually, and cook, stirring constantly until smooth and thickened. Add 2 cups of cheese and stir until melted. Crumble bacon and add to cheese. Spread one-half of the macaroni in buttered baking dish. Arrange one-half of the tomato slices on top. Top with one-half of the cheese sauce. Repeat layers and pour cheese sauce over all. Sprinkle the top with the remaining three-fourths cup of cheese and paprika. Broil four inches from heat, four to five minutes, or until cheese is browned and bubbly. Serves 6.

Variations: Bacon may be omitted and one-half cup chopped ham substituted. Green pepper, celery, and onion, may be added for variety.

Peanut Butter and Honey Snack Balls

¼ cup honey (59 ml)
 1 cup peanut butter (256 g)
 1 cup dry milk (⅓ cup needed for reconstitution) (68 g)

GRAINS, NUTS, AND SEEDS

Mix together the honey and peanut butter. Then add the dry milk and mix well. Make into balls and roll in chopped nuts, sesame seeds, sunflower seeds, wheat germ, oatmeal or crushed, dry unsweetened cereal. Be sure to check label for a peanut butter that doesn't have sugar added to it.

Note: At Easter time, make into the shape of eggs and put in Easter basket.

Cashew-Nut Casserole

2 cups raw cashew nuts (280 g)
2 cups celery, chopped (200 g)
1 cup onions, chopped (120 g)
2 or 3 cans mushrooms, or ½–¾ lb fresh (227 g or 340 g fresh)
3 tablespoons oil (15 ml)
2 cans mushroom soup (10¾ oz/610 g)
2 cans chow mein noodles (5 oz/142 g each)

Roast cashews in oven at 300° F (149° C) about one-half hour. (Watch *carefully* as they brown very fast.) Sauté onion, celery, and mushrooms in about 3 tablespoons oil until transparent. Mix 1½ cans noodles with nuts, vegetables and soup in a low flat ungreased casserole. Sprinkle ½ can noodles over top. Bake ½ hour at 300° F (149° C). Serve immediately. Do not hold. Serves 4–6.

Breads

GENERAL INFORMATION

Through the ages, bread has been called the staff of life. It is the major source of food in many parts of the world today. However, in more developed countries, bread usually is not the main dish but is a marvelous accompaniment to other foods. The aroma and appearance of freshly baked bread—whether it is a yeast bread or a quick bread—contributes much to the pleasure and enjoyment of a meal. So many different ingredients can be added to breads that the variety of tastes and textures seems endless, in addition to the nutritive benefits that can be achieved. The protein quantity and quality of bread can be improved by the addition of some soy flour, nonfat dry milk, eggs, etc.

NUTRITIVE VALUE

Many people have the erroneous idea that bread should be completely omitted from the diet when on a weight reduction plan. A certain amount of bread has its place in any well balanced diet—even one designed for losing weight. Bread contains quite a bit of carbohydrate (mostly in the form of starch) which is important in the production of energy. A slice of white bread provides 13 grams of carbohydrate and 2 grams of protein. The protein contribution may not sound like much at first, but it quickly adds up. Whole grain and enriched breads contribute important amounts of thiamin, riboflavin, niacin, and iron. In addition, whole grain breads contribute a little more of the other B vitamins and the trace minerals. Whole grain breads also contribute fiber or roughage which helps to prevent constipation.

The tortilla is a flat bread used in Mexican foods. It provides the nutrients found in the bread group. On the whole, breads are not a good source of calcium, but the lime treatment given to the corn in the process of making meal known as masa harina provides a great deal of calcium and makes the corn tortilla a good source of calcium. Adding nonfat dry milk in making flour tortillas will improve their protein value, riboflavin, and calcium content.

The chart on page 98 presents a comparison of several flours and grain products used in baking. As you can see, through enrichment, white flour provides more riboflavin than whole wheat flour. The thiamin, niacin, and iron [1] content are about the same for both flours. Whole wheat contains more protein and fat but a little less carbohydrate. However, the whole wheat still contains the undigestible fiber in the bran. It is somewhat less digestible and you may not be able to get all the nutrients which are present.

Soy flour is more concentrated in protein content than wheat flour. About ⅕ the amount of soy flour provides close to the same amount of protein as does white flour. Therefore, small amounts of soy flour can be used to improve both the quantity and quality of the protein in a baked product. *Soy flour does brown more quickly, so watch it!*

COOKING TIPS
1. Put 1 or 2 tablespoons of soy flour in a cup or on the scale and fill the cup to the needed weight with

---

[1] Iron enrichment may be increased.

Table VIII
Nutrient Content of Selected Flours and Grain Products, 100g

Flour or Product	Calories	Protein g	Fat g	Carbohydrate g	Calcium mg	Iron mg	Vitamin A I.U.	Thiamin mg	Riboflavin mg	Niacin mg	Vitamin C mg
White [1]	364	10.5	1.0	76.9	16	2.9	0	.44	.26	3.5	0
Whole wheat	333	13.3	2.0	71.0	41	3.3	0	.55	.12	4.3	0
Soy	326	47.0	0.9	38.0	265	11.1	40	1.09	.34	2.6	0
Wheat germ	363	26.6	10.9	46.7	72	9.4	0	2.01	.68	4.2	0
Wheat bran	213	16.0	4.6	61.9	119	14.9	0	.72	.35	21.0	0
Corn meal [2]	364	7.9	1.2	78.4	6	3.5	440	.44	.26	3.5	0

Values taken from *Composition of Foods*, Agriculture Handbook No. 8, Agricultural Research Service, United States Department of Agriculture, 1963.

[1] Enriched.
[2] Enriched yellow cornmeal (white has little vitamin A value).

regular flour. This will improve the protein value without causing a change in texture and flavor. Soy flour does not contain gluten which gives the structure and volume to wheat flour products.

2. Whole wheat flour can be substituted for white flour. I have obtained satisfactory results in substituting by removing 2 tablespoons of whole wheat flour from each cup which is being substituted.
3. Wheat germ and rice polish [1] can replace 2 tablespoons of flour per cup called for in a recipe.
4. Freshly milled flour has a yellowish color and lacks the qualities necessary to make an elastic, stable dough. When flour is stored for a time, oxidation occurs and the flour matures, becoming whiter and gaining characteristics suitable for baking. This natural process takes several months, requiring storage with the hazards of deterioration and insect and rodent infestation. It was discovered that this natural process could be speeded up by using chemicals (in the form of gas) without removing anything or leaving a significant residue. Unbleached flours, without the chemical treatment, are available in most grocery stores. I do not know how long they may have been permitted to mature naturally. I have used unbleached flour in a great deal of my baking with satisfactory results. In fact, I observed no difference between the flours.

---

[1] Rice polish is the inner bran of rice, containing valuable vitamins and minerals, removed during processing.

Sunday Morning Coffeecake Muffins

¾ cup sugar (140 g)
¼ cup shortening or margarine (47 g)
 1 egg
 3 tablespoons nonfat dry milk
½ cup water (118 g)
⅔ cup flour (80 g)
⅓ cup wholewheat flour (55 g)
 2 tablespoons soy flour
 2 tablespoons wheat germ, untoasted
 2 teaspoons baking powder
½ teaspoon salt
½ cup granola (60 g)

Beat together sugar, shortening, and egg. Mix in the nonfat dry milk and water. Then add the flours, wheat germ, baking powder and salt. Mix well. Spoon into well greased muffin cups. Sprinkle the granola on top. Bake at 400° F (205° C) for 15 to 20 minutes. Makes 12 muffins.

Serve with Tofu Scrambled Eggs (page 41), and half a grapefruit.

Cornbread-Biscuit Stuffing

¼ cup butter or margarine (57 g)
½ onion, chopped fine
 1 stalk celery, chopped (large)
 5 cornbread muffins or pieces, leftover
 5 biscuits, leftover
 1 egg
½ cup walnuts, chopped (50 g)
 1 teaspoon salt
dash pepper
 1 cup chicken broth (235 ml)
 1 teaspoon sage

BREADS 101

Sauté the onion and celery in the butter until tender. Crumble the cornbread and the biscuits. Add the nuts, seasonings, sautéed vegetables, egg, and broth and toss until well blended. Makes about 4 cups stuffing.

Use as stuffing for a pork roast, pork chops, chuck roast, Cornish game hens, chicken breasts, legs, or thighs.

Biscuit Mix

The following two recipes for biscuit mix can be used in a number of recipes which are given and call for a bisquick or biscuit type of base. It is much cheaper to make one's own "bisquick." It must be stored in an airtight and bug-tight container and stored on a cool shelf where it should keep up to eight weeks. If you have extra refrigerator space, store it there and it will keep longer.

If margarine is used in the mix, it gives a distinctive flavor to the products. If polyunsaturated, it is better for those who are on a low cholesterol diet.

Biscuit Mix

    9 cups sifted enriched flour (1125 g or 1.1 kg)
    4 tablespoons baking powder (50 g)
    1 tablespoon salt
1⅓ cups nonfat dry milk (120 g)
1½ cups margarine or shortening (345 g or 285 g)

Put all dry ingredients in large mixing bowl. Cut the margarine or shortening in with pastry blender or two sharp knives until the shortening is in pieces the size of small peas.

Higher Protein Biscuit Mix

- 9 cups sifted enriched flour (1125 g or 1.1 kg)
- 2 cups nonfat dry milk powder (90 g)
- 1 cup soy flour (100 g)
- 5½ tablespoons baking powder (65 g)
- 1½ tablespoons salt
- 1 cup wheat germ (110 g)
- 1½ cups margarine or shortening (345 g or 285 g)

Sift together all dry ingredients except wheat germ. Stir in wheat germ. Cut in margarine or shortening with pastry blender or two sharp knives until shortening is in pieces the size of small peas. This amount stores in two 2 lb. (910 g) coffee cans.

Biscuits

- 2 cups biscuit mix (280 g)
- ¾ cup milk or water (180 ml)

Carefully add the liquid to the biscuit mix to form a stiff dough. Place dough on a lightly floured board and knead about 5 times. Use extra flour if you need to keep the dough from sticking. Then roll or pat out to a ½ in. (1.2 cm) thickness. Cut with a biscuit cutter that is dipped in flour. Bake on an ungreased cookie sheet at 425° F (218° C) for 10 to 15 minutes or until lightly browned. Makes 12 biscuits.

Parmesan Biscuit Bake

- ¼ cup butter (60 g)
- 3 cups biscuit mix (420 g)

¾ cup milk or ¼ cup non-fat dry milk plus ⅔ cup water (235 ml)
¼ cup Parmesan cheese, grated (25 g)
2 tablespoons sesame seeds, toasted

Melt butter or margarine and set aside. Stir together biscuit mix and milk to make a soft dough. Gently form dough into a ball on floured surface and knead 5 times. Roll out to one-fourth inch (.6 cm) thickness and cut with 2½ inch (5.2 cm) biscuit cutter. Combine Parmesan and sesame seeds in shallow dish. Dip each biscuit in melted butter then in Parmesan. Stand on edge in two rows in greased loaf pan. Slightly overlap biscuits in the center. Sprinkle with remaining cheese mixture and sesame seeds. Bake at 450° F (177° C) for 15 to 20 minutes.

Quick Cheese Bread

3¾ cups biscuit mix (455 g)
1 egg
1¼ cups milk (284 ml) or ½ cup (40 g) dry instant milk plus 1 cup (237 ml) water
½ teaspoon dry mustard
1¼ cups sharp cheese—add after last blending (115 g)
sesame, poppy, or caraway seeds

Blend the biscuit mix, egg, milk, and mustard for one minute. Then mix in cheese by hand. Top with sesame, poppy, or caraway seeds. Bake at 350° F (177° C) for about 30 minutes in a loaf pan.

Butter Top Coffee Cake

2 cups biscuit mix (280 g)
2 tablespoons sugar
1 egg
⅔ cup water or milk (156 ml)

Mix ingredients and spread dough in 2 round cake pans or 1 large rectangular pan. The dough should be thin. Butter the top of the dough and sprinkle with topping (see below). Then, with finger, punch 15 to 20 holes on top of dough. Bake at 400° F (205° C) for about 15 minutes.

Topping

¾ cup powdered sugar (75 g)
3 tablespoons flour or biscuit mix
½ cup butter or margarine (115 g)

Whole Wheat Muffins

1 cup all-purpose flour (123 g)
1 cup whole wheat flour, less two tablespoons (140 g)
3 tablespoons sugar
1 tablespoon baking powder
½ teaspoon salt
½ teaspoon cinnamon
¼ teaspoon nutmeg
2 egg whites or 1 large egg
¼ cup margarine or oil, melted (59 ml)
1 cup milk (235 ml)

Sift or stir the dry ingredients togther. Make a well in

the center. Beat egg whites and combine with oil and milk. Pour at once into flour well. Stir just to moisten ingredients, about 15 strokes. Bake at 425° F (218° C) for about 25 minutes.

Variation: Add ⅔ cup finely chopped walnuts.

East Texas Cornbread

    2 eggs
    2 cups buttermilk (470 ml)
    1 teaspoon soda
1½ cups corn meal (225 g)
    ¼ cup flour, white or whole wheat (31 g)
    ¼ cup soy flour (25 g)
    1 teaspoon salt

Beat together the eggs and buttermilk. Stir together the dry ingredients. Add the liquid mixture to the dry ingredients. Mix until smooth. Pour into a hot, oiled, square pan or muffin pan. Oiling a cast iron skillet with bacon drippings works very well for baking the cornbread. Bake at 450° F (232° C). Muffins require 10 to 15 minutes to bake and the bread requires 20 to 25 minutes. Makes 12 muffins or pieces.

Store leftover cornbread in the freezer and use later in Cornbread-Biscuit Stuffing (page 100).

This recipe produces a compact texture and somewhat heavy cornbread. However, it is moist and very tasty. It has been passed down from my husband's grandmother, with my addition of the wheat and soy flour.

Persimmon Bread

   1 cup persimmon pulp (250 g)
   2 eggs
   1 cup sugar (187 g)
2½ cups flour (308 g)
   3 teaspoons baking powder
½ teaspoon nutmeg
   1 teaspoon cinnamon
½ teaspoon salt
½ cup oil (118 ml)
   1 teaspoon vanilla (5 ml)
½ cup walnuts, chopped (50 g)

In the bowl of an electric mixer, combine persimmon pulp, eggs, and sugar, beating well. Stir together flour, baking powder, nutmeg, cinnamon, and salt. Add sifted dry ingredients to the persimmon mixture with oil.

Blend well. Stir in vanilla. Fold in chopped nuts. Turn into a greased loaf pan and bake at 325° F (163° C) for one hour, or until tests done. Serves 8.

Pumpkin Bread

3½ cups flour (use part soy flour) (430 g)
   2 teaspoons soda
1½ teaspoon cinnamon
   1 teaspoon nutmeg
1½ teaspoon salt
   3 cups sugar (558 g)
   1 cup oil (235 ml)
   4 eggs, beaten
⅔ cup water (156 ml)

2 cups pumpkin, canned or fresh (456 g)
1 cup walnuts, chopped (100 g)

Sift all dry ingredients in a large bowl and make a hole in center. Combine the oil, eggs, water, pumpkin, and nuts and add to the dry ingredients. Mix well with beater until creamy. Pour into three well greased one-pound coffee cans or two loaf pans. Do not flour. Bake for one hour at 350° F (177° C) or until toothpick comes out clean.

Quick Brown Bread

2 tablespoons butter or margarine
½ cup sugar (93 g)
½ cup molasses (118 ml)
1 cup buttermilk (235 ml)
1 egg
1 cup whole wheat flour (160 g)
1 cup all purpose flour (123 g)
2 tablespoons soy flour
1 teaspoon *each* salt and soda
½ teaspoon *each* cinnamon and ginger

Cream butter and sugar until creamy. Add molasses, buttermilk, and egg. Beat well. Add flours, salt, soda, and spices. Beat 2 minutes. Stir in raisins. Spoon into a well greased loaf pan. Bake at 350° F (177° C) for about 60 minutes.

Variation: Add ½ cup raisins (83 g).

Lettuce Bread

- 1 cup lettuce, shredded (50 g)
- 2 tablespoons wheat germ
- 2 tablespoons soy flour
- 1½ cups flour, sifted (185 g)
- 2 teaspoons baking powder
- ½ teaspoon baking soda
- 1 teaspoon salt
- ⅛ teaspoon *each* ground mace, ginger, cinnamon
- 1 cup sugar or 1 cup packed brown sugar (187 g or 210 g)
- ½ cup oil (118 ml)
- 1 teaspoon lemon rind
- ½ cup chopped walnuts or ⅔ cup mayonnaise (30 g or 156 ml)
- 2 eggs, beaten

Core, rinse, and drain lettuce; chill until crisp in airtight container. Chop lettuce. Sift flour with baking powder, soda, salt, wheat germ, and spices. Combine sugar, oil, and lemon rind; mix in flour combination and chopped lettuce. Add eggs and beat well. Stir in walnuts. Turn into greased and floured loaf pan. Bake at 350° F (177° C) for 55 minutes. Cool in pan; invert on wire rack. Sprinkle with powdered sugar.

Pineapple-Cheese-Nut Loaf

- 1 can crushed pineapple (13¼ oz/371 g)
- 1 cup ready-to-eat bran cereal (80 g)
- 2⅛ cups sifted all-purpose flour (277 g)
- 2 tablespoons soy flour
- ⅓ cup sugar (69 g)
- 3 teaspoons baking powder
- 1 teaspoon salt
- ½ teaspoon soda

1 cup sharp Cheddar cheese, grated (92 g)
2 eggs, beaten
3 tablespoons soft or melted shortening (45 ml)
½ cup walnuts, chopped, preferably black (50 g)

Combine undrained pineapple and bran cereal; let stand while measuring remaining ingredients. Resift flours with sugar, baking powder, salt, and soda. Add cheese and mix lightly. Combine eggs and shortening with pineapple mixture. Add sifted dry ingredients and walnuts. Mix just until all of flour mixture is moistened. Spoon into well-greased loaf bread pan. Bake on lowest rack of moderately low oven, 325° F (163° C) about one hour and 10 minutes until loaf tests done. Let stand 10 minutes then invert onto wire rack to cool. Makes one large loaf or two smaller loaves.

Wheat Germ Banana Nut Loaf

2 tablespoons soy flour
2 cups flour (246 g)
½ cup wheat germ (55 g)
½ cup sugar (100 g)
3 tablespoons baking powder
1 teaspoon salt
½ cup walnuts, chopped or sunflower seeds (50 g)
1 cup mashed banana (2 large) (300 g)
¼ cup dry milk (118 ml)
¼ cup salad oil or melted butter (59 ml)
1 egg
½ cup water (118 ml)

Mix all dry ingredients. Combine banana, milk, shortening, and egg. Mix well with dry ingredients. Spread in greased loaf pan. Bake 350° F (177° C) for 50 to 60 minutes. Remove from pan immediately and cool.

Tortillas (Flour)

    3 cups flour (369 g)
1½ teaspoons salt
 ¾ teaspoon baking power
 ¼ cup shortening (47 g)
 ⅓ cup nonfat dry milk (30 g)
    1 cup water (237 ml)

Mix dry ingredients and cut in shortening. Mix thoroughly as when making pie crust, then add liquid. Knead dough with hands as when making yeast bread; when dough is smooth, pinch off dough the size of a golf ball. Place balls on wax paper, oil tops, cover and let stand 15 minutes. (The longer you let the dough rest in refrigerator or on drain board, the easier it is to work with. As long as the dough is oiled well you can keep it 24 hours in refrigerator.) Roll out and cook on hot cast iron griddle or fry pan. Makes about 18 tortillas. Use as a hot bread or serve with beans or meat. Also good for enchiladas.

Sopaipillas (A Fried Bread)

    2 cups flour (246 g)
 ¾ teaspoon salt
 ½ teaspoon baking powder
1½ teaspoons shortening
 ¼ cup warm water (adding more water as needed) (59 ml)

Combine dry ingredients and cut in shortening. Make a well in center of dry ingredients. Add liquid to dry ingredients and work into dough, adding only enough liquid to make a firm dough. Knead dough 15 to 20 times and let rest for 10 minutes. Roll dough to ⅛ inch (.3 cm) thickness and then cut in squares or triangles. Cover the dough with a towel. Fry in very hot fat, about 420° F (218° C). They should puff very soon after being dropped into the hot fat. Fry only a few at a time.

To serve as a main dish, fill with a meat mixture, cheese, beans, onions, yogurt, sour cream, lettuce, etc. Dip in honey to serve as a dessert.

Pocket Bread

   1 cup whole wheat flour (160 g)
1½ to 2 cups flour (185 g to 246 g)
   1 teaspoon sugar
   1 teaspoon salt
   1 package dry yeast
   1 cup warm water (235 ml)

In a large bowl mix 1 cup mixed flour, sugar, salt, and yeast. Gradually add the water to the dry ingredients and beat well. Gradually add flour until a soft dough is reached; then turn out on a lightly floured board and knead until smooth and elastic. Place in a greased bowl, turning to grease top. Cover with a warm damp cloth and let rise in a warm place until double. Then punch down dough and let rest for 30 minutes on the lightly floured board. Preheat over to 450° F (230° C). It is important that the oven is *very* hot when you put the bread in. Divide dough into 6 equal pieces; shape into a ball and roll out in a circle to a six-inch (15 cm) diameter. Bake on lowest rack of oven for 5 minutes and on center rack for 3–5 minutes or until lightly browned and centers puff up. Split after cool.

High Protein Bread

2 packages dry yeast
¼ cup honey (60 ml)
2 cups milk (470 ml)
½ cup margarine (115 g)
3 teaspoons salt
3 eggs
½ cup wheat germ (55 g)
1 cup flour (125 g)
1 cup soy flour (100 g)
5 cups whole wheat flour (800 g)

Dissolve yeast in ¼ cup (60 ml) warm water with 1 teaspoon honey. Add yeast mixture to large bowl containing milk, margarine, honey, eggs, and salt. Add wheat germ, white flour, and soy flour and beat well. Then mix in whole wheat flour gradually. Knead bread well until smooth. Allow to rise about 1 hour or until double in bulk. Punch down and shape into loaves. Place loaves in greased pans and allow to rise about 40 minutes or double in size. Bake at 400° F (205° C) for the first 10 minutes; then bake at 350° F (177° C) for 20 to 30 minutes.

This bread provides complete protein and is quite satisfying. It is good toasted, especially for breakfast.

Hi-Protein Vegetable Bread

1 can evaporated milk (13 fluid oz/282 ml)
2 carrots, cut in large pieces
2 stalks of celery, cut in large pieces
4 parsley flowerets
¼ head of cabbage, cut in large pieces
4 eggs
2 tablespoons honey (30 ml)
2 tablespoons margarine

2 teaspoons sugar
2 packages yeast
¾ cup warm water (177 ml)
2 cups unbleached all-purpose flour (246 g)
¼ cup soy flour (25 g)
about 7 cups whole wheat flour (1,120 g)

Put milk, vegetables, eggs, honey, and margarine in a blender and blend well. Combine sugar, yeast, and water, and let rise for about 10 minutes. Then add the yeast mixture to the vegetable mixture. Stir in the all-purpose flour and soy flour. Beat well. Then add the whole wheat flour gradually until it can be kneaded. Knead until smooth and elastic. Place in a greased bowl and cover with damp cloth. Let rise until double in size. Punch down and shape into 4 loaves. Place in greased bread pans and permit to rise until double in size. Bake at 425° F (218° C) for 10 minutes and then reduce heat to 325° F (163° C) for 20 minutes or until done. Brush with milk to give a soft crust.

Carrot Brownies

½ cup butter or margarine (114 g)
1½ cups brown sugar (315 g)
2 eggs
2 cups flour (part soy) (246 g)
2 teaspoons baking powder
½ teaspoon salt
2 cups carrots, finely grated (200 g)
½ cup walnuts, chopped (50 g)

Melt butter; add sugar and stir until well blended. Remove from heat and beat in eggs. Beat in all remaining ingredients except nuts. Pour mixture into 2 square pans or 1 large rectangular shaped pan. Sprinkle with the chopped nuts. Bake at 350° F (177° C) for 30 minutes.

Orange Nut Bread

1½ cups flour (184 g)
½ cup whole wheat flour (80 g)
2 tablespoons wheat germ
2 tablespoons soy flour
¾ cup sugar or brown sugar (79 g)
¾ cup walnuts, chopped (75 g)
1 egg, beaten
2 tablespoons oil (30 ml)
¾ cup orange juice (177 ml)
1 tablespoon grated orange rind (optional)

Stir or sift dry ingredients together; add nuts. Combine egg, oil, orange juice, and rind. Pour liquid ingredients into flour mixture; stir only until smooth. Bake in oiled loaf pan for 60 minutes at 350° F (177° C).

Cheese Sourdough French Bread

1 package chive cream cheese, softened (3 oz/84 g)
2 tablespoons butter or margarine, softened
1 teaspoon prepared horseradish
1 large loaf sourdough French bread

Combine cream cheese, butter or margarine, and horseradish. Bias-cut each French loaf into slices, cutting to but not through bottom crust. Spread butter mixture between slices. Bake on baking sheet in 400° F (205° C) oven for 12 to 15 minutes or until browned.

Variation: You can substitute English Muffin halves for the slices of French bread.

Pastry Shell (8 in/20 cm)

1 tablespoon soy flour

½ cup all-purpose flour, or whole wheat pastry flour (60 g)
½ teaspoon salt
¼ teaspoon baking powder
¼ cup margarine (55 g)

Combine first four ingredients in a mixing bowl. Blend in the margarine with a pastry blender until mixture leaves side of bowl. Shape into ball; wrap in foil; chill one hour. Roll out on lightly floured pastry cloth with a stockinette-covered rolling pin. Roll to fit in (20 cm) pie plate. Turn pastry under; flute edge.

Pastry Shell (9 in/22.5 cm)

1 cup less 1 tablespoon flour (115 g)
1 tablespoon soy flour
½ teaspoon salt
⅓ cup shortening (65 g)
1 tablespoon shortening
2 tablespoons ice water (30 ml)

Stir together flours and salt. Cut ⅓ cup (65 g) shortening into the flour with a pastry blender. It should look like a coarse meal when finished. Cut the remaining 1 tablespoon shortening into the mixture with 2 knives. The pieces should look like peas. Sprinkle in the water and mix with a fork until all the flour is moistened. Gather dough together and press firmly into a ball. Roll out on lightly floured pastry cloth with stockinette-covered rolling pin. Roll to fit 9 in. (22.5 cm) pie plate. Turn pastry under; flute edge.

If a shell is to be baked prior to filling, prick the bottom of the crust with a fork, and bake at 475° F (246° C) for 8 to 10 minutes.

Eggs and Cheese

GENERAL INFORMATION

The recipes in this section utilize eggs, cheese, and other milk products as the principal sources of protein. In addition to cheese, yogurt and nonfat dry milk are specifically included. Eggs and cheese are particularly good complements in recipes using grain products and vegetables such as squash, eggplant, chiles, and spinach. These vegetables are not very high in protein content but do provide a tasty foundation for an entree using eggs and cheese.

NUTRITIVE VALUE

Both egg protein and milk protein are of excellent quality. Cheese contains less water than milk; therefore its protein content is more concentrated. Milk products are also good sources of B vitamins (especially riboflavin), vitamins A and D (when fortified), calcium, and phosphorus. They are not good sources of iron or vitamin C. Eggs also lack vitamin C but contain iron and most of the other minerals and vitamins. A drawback in using eggs to supply protein is the fact that they are rich in cholesterol. In low cholesterol diets it is recommended that the use of cholesterol rich foods be restricted, but these recipes can be used when calculated into the weekly allotment of eggs.

Table IX
Nutrient Content of Selected Foods, 100g [1]

Approx. Measure	Food	Calories	Protein g	Fat g	Carbohydrate g	Calcium mg	Iron mg	Vitamin A I.U.	Thiamin mg	Riboflavin mg	Niacin mg	Ascorbic Acid mg
2	Egg	163	12.9	11.5	0.9	54	2.3	1,180	.11	.30	.1	0
3½ oz.	Cheese, Cheddar	398	25.5	32.2	2.1	750	1.0	1,310	.03	.46	.1	0
2/5 c.	Milk, whole	65	3.5	3.5	4.9	118	trace	140	.03	.17	.1	7
1 3/5 c.	Milk, nonfat dry	363	35.9	0.8	52.3	1308	.6	30	.35	1.80	.9	0
3/5 c.	Sour Cream	200	trace	16.0	8.0	96	trace	759	.03	.13	trace	trace
2/5 c.	Yogurt	50	3.4	1.7	5.2	120	trace	70	.04	.18	.1	1
1/2 c.	Eggplant, cooked	19	1.0	.2	4.1	11	0.6	10	.05	.04	.5	3
1/2 c.	Mushrooms, canned	17	1.9	.1	2.4	6	0.5	trace	.02	.25	2.0	trace
1/2 c.	Zucchini, cooked	12	1.0	.1	2.5	25	0.4	300	.05	.08	.8	9

[1] Adapted from *Composition of Foods*, Agriculture Handbook No. 8, Agricultural Research Service, United States Department of Agriculture, 1963.

COOKING TIPS

1. For low cholesterol diets, egg whites can be substituted for whole eggs in some of the recipes. An egg substitute may also be used in place of the eggs called for. Skim or nonfat milk and its products should be used, and they are usually cheaper.
2. Use low to moderate temperatures; do not cook too long a time.
3. Substitute yogurt for sour cream in recipes: yogurt is cheaper, less caloric, and contains more protein.
4. Boost the protein value of many recipes by adding a couple of tablespoons of nonfat dry milk.
5. Make your own low-fat milk by mixing equal parts of reconstituted nonfat dry and whole milk.

Yogurt

- 3 cups powdered milk (207 g)
- 1 large can evaporated milk (13 oz/382 ml)
- 6 cups water (1410 ml or 1.4 l)
- 3 tablespoons yogurt (either purchased or from last batch)
- 1 tablespoon sugar
- ½ teaspoon unflavored gelatin

Soften gelatin and add boiling water to make 1 cup. Add sugar. Preheat oven to 250–300° F (121° to 149° C) while you mix remaining ingredients in heat-proof bowl. Cover, place in oven and turn oven OFF. Leave 8 to 10 hours or overnight. Makes 2 quarts (2 l).

A few drops of vanilla adds flavor if you are going to add fruit.

I suggest letting the yogurt set up in either cup or pint jars rather than letting it set in the oven in the large bowl. This recipe does not get as firm as commercial yogurt. Since it is plain, it can be used in any recipe calling for yogurt.

Yogurt and Jello Salad

1 package fruit flavored jello
1 cup boiling water (235 ml)
1 cup cold water (235 ml)
1 cup container fruit flavored yogurt, your own or commercial (227 g)

Make jello in usual manner. Refrigerate until cold. Add the fruit flavored yogurt. Whip until blended. Pour into mold, and chill.

Yogurt Pie

2 containers of plain yogurt, your own or commercial (8 oz/227 g each)
1 one pound container of small curd cream style cottage cheese (454 g)
4½ teaspoons unflavored gelatin
½ cup milk (118 ml)
½ cup sugar (93 g)
1 teaspoon vanilla (5 ml)
1 nine-inch graham cracker crust (22.5 cm)
nutmeg

Blend the cottage cheese and one container or cup of yogurt together in blender. Soften the gelatin in the milk, then dissolve over hot water, stirring constantly. Stir into cheese mixture. Add remaining container of yogurt, sugar, and vanilla. Stir until well mixed and pour into graham cracker crust. Sprinkle with nutmeg. Chill until set (about 3 or 4 hours). Serves 6 to 8.

Graham Cracker Crust

9 double graham crackers, crushed
2 tablespoons sugar
⅓ cup butter or margarine, melted (90 g)

Stir all ingredients together and press into pie pan.

Yogurt Fresh Vegetable Dip

1 cup yogurt (8 oz/227 g)
1 cup mayonnaise or salad dressing (235 g)
1 teaspoon Beau Monde
1 teaspoon dried onion, minced
1 teaspoon dill weed
1 teaspoon chives, dried
salt and pepper to taste

Mix all ingredients and let set overnight in the refrigerator to thoroughly blend flavors.

Variation: Substitute sour cream for the yogurt and use fresh chopped chives.

Yogurt Banana Dessert

4 firm bananas
⅔ cup yogurt (150 g)
⅓ cup brown sugar, packed (55 g)
1 lemon, juiced

Peel bananas and cut in half lengthwise. Place in shallow baking dish. Sprinkle lemon juice over the bananas. Then spoon yogurt over bananas. Sprinkle with brown sugar. Place under broiler and broil 3 to 5 minutes, until bananas are just tender. Makes 4 servings.

Bananas are rich in potassium, so this recipe provides a lot of potassium.

Variation: Substitute sour cream for the yogurt. However, the protein value will be less.

Chile Rellenos

A favorite recipe from the American Southwest.

1 can green chiles or 6 fresh chiles (10 oz/284 g)
sufficient Monterey Jack cheese to fill chiles (½ × ¾"/
    12 cm by 18 cm)

2 eggs
2 tablespoons flour

Rinse and remove seeds from the chiles. Slit one side and fill with one strip of cheese. Roll each in flour. Beat egg whites until stiff. Beat egg yolks and add 2 tablespoons flour. Fold into egg whites. Roll chiles in flour and then in batter. Fry both sides in small amount of oil. Put into sauce and serve. Allow 2 chiles per person.

Sauce

½ onion, chopped fine
1 clove garlic, chopped fine
1 tablespoon oil (15 ml)
1 teaspoon salt
½ teaspoon pepper
1 teaspoon marjoram or oregano (marjoram is best)
1 cup chicken broth or 1 bouillon cube plus 1 cup water (235 ml)
1 can tomatoes, stewed (454 g)

Sauté the onion and garlic in oil. Combine all ingredients and simmer for at least 10 minutes. This sauce is also good on eggs or tofu patties.

Chilaquilas

Tortillas
Monterey Jack cheese
2 eggs
tomato sauce (from Chile Rellenos)
sour cream

Cut tortillas in fourths, forming triangles. Put cheese between two pieces. Beat egg whites until stiff and add egg yolks. Dip tortilla with cheese into egg mixture and fry on both sides. Make tomato sauce as for chili rellenos. simmer tortillas in sauce. Depending on how many you make, you may have to double tomato sauce recipe. If desired, add salt to cheese. Serve with sour cream on top.

Huevos Rancheros (Ranch Style Eggs)

- ¼ cup onion, chopped (40 g)
- 1 clove garlic, chopped
- 3 tablespoons butter
- 3 large fresh tomatoes, chopped
- 1½ tablespoons salsa jalapena, Picante Sauce, or taco sauce
- 8 tortillas
- 8 eggs

Sauté the onion, garlic, and butter. Add the tomatoes and salsa jalapena and simmer for 30 minutes.

Fry tortillas and drain, fry eggs and place one on each tortilla. Top with vegetable mixture. Serves 8.

Variation: Spread beans over each tortilla. Soft boil the eggs (3 minutes from the time they come to a boil).

Huevos Reuveltos (Green Chile Omelet)

- 2 tablespoons butter
- 8 eggs
- 2 tablespoons cream (30 ml)
- 2 small tomatoes, peeled and chopped (fresh)
- 2 tablespoons green chiles, chopped
- 2 tablespoons onions, chopped
- 2 tablespoons parsley, chopped

⅔ cup mild Cheddar cheese, cubed (60 g)
salt and pepper to taste

Melt butter. Beat eggs with cream and pour into pan. As eggs begin to set add all the other ingredients and scramble. Serve with juice and refried beans. Serves 6–8.

Green Chile and Cheese Casserole

1 medium onion, finely chopped
2 tablespoons butter or margarine
1 can tomato sauce (8 oz/227 g)
1 can green chile peppers, chopped (4 oz/113 g)
½ teaspoon salt
2 eggs, slightly beaten
1 cup milk (235 ml)
½ lb. Monterey Jack cheese cut in small cubes (227 g)
½ cup Cheddar cheese, shredded (46 g)
paprika
1 cup sour cream or yogurt (227 g)
1 package corn chips (11 oz/308 g)

Sauté onion in butter or margarine until transparent. Stir in tomato sauce, peppers, and salt. Simmer 5 minutes. Remove from heat. Combine eggs and milk, stir into sauce. Place half the corn chips in a two quart casserole. Add, in layers, half the Monterey Jack cheese and half the sauce. Repeat and top with sour cream. Sprinkle with shredded Cheddar cheese and paprika. Bake uncovered in 350° F (177° C) oven 30 minutes. Serves 6.

Egg Fu Yung

1 lb. bean sprouts (454 g)
1 can cocktail shrimp (126 g)
4 large eggs
1 can mushrooms (56 g)
½ onion, chopped
2 green onions, chopped
½ teaspoon monosodium glutamate (optional)
½ teaspoon salt
oil for pan frying

Sauce

2 beef bouillion cubes
2 tablespoons flour
½ cup water plus liquid from can of mushrooms (118 ml + liquid)
1 small can of mushrooms, fried (56 g)
pinch of monosodium glutamate (optional)

Add the seasonings to the eggs. Add shrimp. Mix onions with the bean sprouts and mushrooms and then mix with the egg mixture. Fry in oil. Stack like pancakes and pour sauce on top. Serve with soy sauce. Serves 6–8.

Banana Pudding or Pie Filling

1 banana, medium size
½ cup sugar (120 g)
⅓ cup flour (40 g)
1 egg
1 egg yolk
½ teaspoon salt
2 tablespoons butter or margarine

EGGS AND CHEESE

⅔ cup nonfat dry milk (60 g)
1½ cups water (345 ml)
1 teaspoon vanilla (5 ml)

Beat together the sugar, egg and egg yolk. Then beat in the flour. Stir in as much of the dried milk as the mixture will hold. Blend in the water and add the remainder of the dried milk. Add the butter. Cook over medium heat or preferably in a double boiler to prevent scorching. Cook about a minute after it begins to boil gently and thicken slightly. Remove from heat and stir in vanilla. Slice the banana on the bottom of the bowl and stir the pudding into the bananas. Serves 4.

Quiche Lorraine

1 unbaked pie shell (8 in/20 cm)
2 eggs, uncooked, beaten slightly
½ cup lean ham, chopped (80 g)
1 cup cheese, shredded—combination of Swiss and Cheddar (92 g)
1 can evaporated skim milk (13 oz/382 ml)
2 teaspoons parsley flakes
½ teaspoon salt
pepper

Combine eggs, ham, cheese, milk, parsley, and salt and pepper, and pour into the pastry shell (page 115). Bake in 400° F (205° C) oven 10 minutes. Reduce heat to 350° F (177° C) and bake 20 minutes or until knife inserted comes out clean. Cool 10 minutes.

Corn Quiche Lorraine

1 unbaked pastry shell (9 in/22.5 cm)
7 bacon slices
½ onion, chopped fine
⅓ cup Swiss cheese, shredded (30 g)
⅔ cup Cheddar cheese, shredded (60 g)
1 tablespoon flour
3 eggs, slightly beaten
1 can cream style corn (1 lb, 1 oz/482 g)
½ cup evaporated milk (118 ml)
dash fresh-ground pepper
dash cayenne

Prick sides and bottom of pastry shell (page 115). Bake at 450° F (232° C) for 8 minutes. Cook bacon until crisp. Remove from skillet; drain and crumble. Sauté onion in one tablespoon bacon drippings until tender. Toss crumbled bacon and cheeses with flour; sprinkle evenly in pastry shell. Blend together eggs, onion, corn, evaporated milk, pepper, and cayenne. Pour mixture over bacon and cheese. Bake at 350° F (177° C) 50 minutes, or until knife inserted in center comes out clean. Garnish with crisply cooked crumbled bacon, if desired. Cool 10 minutes before cutting. Serves 6–8.

Cheese Soufflé

4 tablespoons butter or margarine (60 g)
1 tablespoon bread crumbs

3 tablespoons flour
½ teaspoon salt
⅛ teaspoon pepper
⅛ teaspoon ground nutmeg
¾ cup milk, hot (177 ml)
⅔ cup Swiss cheese, grated (62 g)
5 egg whites
4 egg yolks

Preheat oven to 400° F (205° C). Use 1 tablespoon of butter to grease the inside of a 6-cup (1,410 ml) soufflé dish or deep casserole. Sprinkle the bottom and sides with the bread crumbs. Melt remaining butter and stir in flour, salt, pepper, and nutmeg and cook, stirring constantly, just until mixture bubbles. Stir in milk slowly and continue cooking until sauce thickens. Remove from heat and stir in all but 2 tablespoons cheese. Stir only until cheese melts. While cheese mixture is cooling, beat egg whites until they form soft peaks in a medium sized bowl. Beat egg yolks in a large bowl until thick and fluffy. Beat in cheese mixture, a small amount at a time. Then beat in about 2 tablespoons of the egg whites. Fold in the remaining egg whites. Spoon carefully into soufflé dish. Sprinkle with remaining cheese. Lower heat to 375° F (190° C) and bake for 35 minutes or until golden brown and firm on top and a knife inserted in the center comes out clean.

Variations: Add a package of frozen broccoli or spinach, slightly cooked, to cheese mixture. Also, add ½ cup (80 g) chopped ham or **Spam**.

Eggplant Parmesan

1 medium eggplant, sliced lengthwise
Tomato sauce, recipe below
½ pound *Italian* cheese, sliced (227 g)

Place sliced eggplant in colander—a layer of eggplant and salt, alternately. Put a weight of some kind on top and let set in a place to drain for a couple of hours. Rinse salt from eggplant.

Tomato Sauce
1 can tomatoes (1 lb/454 g)
1 can tomato paste (6 oz/170 g)
1 medium onion, sliced
salt and pepper to taste

Cook slowly for about 45 minutes. Place eggplant slices, sliced *Italian* cheese and tomato sauce in casserole alternately, until all is used. Bake in moderate oven (350° F/177° C) approximately 30 minutes.

Cheese Chops

2 eggs
⅓ cup milk (78 ml)
1 tablespoon butter
1 cup cracker crumbs (59 g)
salt, pepper, and mustard to taste
½ pound Cheddar cheese, grated (227 g)

Mix ingredients together and shape into patties. Chill 2 hours or overnight. Coat with cracker crumbs. Brown in oil in skillet and serve at once. Serves 3.

Cheese, Chive, and Egg Stuffed Eggplant

1 large eggplant
¼ cup oil (59 ml)
2 cloves garlic, chopped

1 can tomatoes, chopped (1 lb/454 g)
¼ cup freeze-dried or frozen chives, chopped (60 ml)
½ cup halved black olives, pited (70 g)
4 hard cooked eggs, chopped
1 cup seasoned dry bread crumbs (50 g)
½ lb. sharp or mild Cheddar cheese, coarsely grated (8 oz/227 g)

Cut eggplant in half, lengthwise. Scoop out eggplant, leaving a shell ¾ in. thick. Chop the eggplant which has been removed. In a large skillet heat oil and add chopped eggplant and garlic. Simmer until eggplant is wilted. Stir in tomatoes, chives, and olives and simmer 5 minutes. Stir in eggs and crumbs. Fold in half of the cheese. Use mixture to stuff eggplant shells. Sprinkle top with cheese. Bake at 350° F (177° C) for 35 to 40 minutes or until top is brown. Each eggplant half serves two.

Gold and Green Casserole

3 medium size zucchini
½ teaspoon salt
½ cup water (118 ml)
2 eggs, beaten
¼ teaspoon salt
⅛ teaspoon pepper
1 teaspoon onion, grated
1 12 oz. can corn, drained, or one package of frozen corn (336 g)
2 cups Cheddar cheese, grated (184 g)

Wash and slice zucchini into half-inch pieces. Boil in salted water about 10 minutes or until tender. Drain well and mash. Drain again. Mix eggs, salt, pepper, and onion. Combine egg mixture and corn with zucchini and cheese. Pour into greased casserole. Top with remaining cheese. Bake at 350° F (177° C) for 30 to 40 minutes or until set. Serves 4–5.

Puffer (Potato Pancakes)

4 to 5 medium potatoes
3 eggs, beaten
1 teaspoon onion, grated
1 tablespoon flour
4 to 5 drops Tabasco sauce (optional)
salt to taste
oil for pan frying

Peel and grate the potatoes. Combine with other ingredients and fry in a skillet or on a griddle in a considerable amount of oil. Serves 4–5.

Serve with canned fruit such as applesauce, peaches, or apricots. You can also serve with syrup.

Note: When I was a little girl, I used to look forward to Sunday night supper at one of my aunt's house. She usually fixed these pancakes and I thought they were the greatest. At the time I didn't realize they were a source of protein and really didn't care. They were good and that's all that mattered. These pancakes reflected my aunt's German heritage and a method of getting protein without using meat.

Squash Casserole

1½ lbs. zucchini or crook-neck squash (680 g)
  1 onion, chopped fine
 ¼ cup margarine or butter (55 g)
  1 cup Cheddar cheese, grated (90 g)
  1 teaspoon monosodium glutamate (optional)
  1 teaspoon salt
 ⅛ teaspoon pepper
  2 eggs, beaten
1½ cups soft bread crumbs (75 g)
  2 tablespoons margarine or butter

Cook whole squash until almost done. Cool. Sauté onions in margarine. Cut squash in cubes and add to onions. Stir in the cheese, seasonings, and eggs. Spoon into buttered casserole. Mix the 2 tablespoons of margarine with the bread crumbs. Top the casserole with the crumb mixture and bake at 350° F (177° C) for 30 to 45 minutes. Serves 4.

Meat Stretchers

GENERAL INFORMATION

The word "meat" is used to denote the flesh of the warm-blooded animals we are familiar with such as beef, pork, chicken, etc. Many of the recipes using meat are based on recipes from other countries where people have had to be most clever in making scarce and expensive meat go farther. Large quantities of vegetables stretch the meat, or the meat may be cut up into a number of small pieces. It may then be served with a grain such as rice or some other staple. An extra bonus is that many of the recipes require a short cooking time, helping to conserve energy.

NUTRITIVE VALUE

Since the red meats contain quite a bit of fat and cholesterol, it may be to our advantage to somewhat decrease our meat intake. However, I do not recommend completely excluding meat from the diet because, in addition to containing a relatively high percentage of good quality protein, it is one of our best sources of available iron and zinc, as well as other minerals. We need to ingest 10 to 18 mg. of iron daily. It also provides a number of the B vitamins: in fact, vitamin B_{12} is found only in animal food sources. I must point out that meat is deficient in calcium and vitamin C. The table on page 133 gives a partial nutritive analysis of some meats used. One hundred grams equals approximately 3½ ounces.

Table X
Nutrient Composition of Some Cooked Meats, 100 Grams [1]

Meat	Cal.	Protein	Fat	Carbo-hydrate	Cal-cium	Iron	Vita-min A	Thia-min	Ribo-flavin	Nia-cin	Ascorbic Acid
		g	g	g	mg	mg	I.U.	mg	mg	mg	mg
Beef, flank	196	30.5	7.3	0	14	3.8	10	.06	.23	4.6	0
Beef, chuck	327	26.0	23.9	0	11	3.3	40	.05	.20	4.0	0
Pork, loin	387	23.5	31.8	0	10	3.1	0	.88	.25	5.3	0
Chicken	248	27.1	14.7	0	11	1.8	420	.08	.14	8.2	0

[1] Values taken from *Composition of Foods*, Agriculture Handbook No. 8, Agricultural Research Service, United States Department of Agriculture, 1963.

Except for flank steak, the red meats contain more saturated fat than chicken or turkey, and the use of poultry is encouraged on low calorie diets and low cholesterol diets. Both chicken and turkey are cheaper than the red meats today.

COOKING TIPS
1. Buy a large 7-bone chuck roast and cut it into small steaks, stew, squares for shish kebab, chile meat, thin strips for teriyaki and skewered beef roll-ups, or cut into a roast for pot roasting or stuffing. Use bones for soup or beef broth.
2. Cook quickly if broiling or pan frying and the meat will be tender without using a tenderizer. The longer it is broiled the tougher it becomes.
3. Learn to cut up your own chicken to reduce cost.
4. Bone wings, legs, and thighs, and stuff to make the meat go farther.
5. Learn to include ground turkey, turkey tenderloins (from breast), wings, thighs, legs, etc. to reduce cost. During the beef shortage and when beef prices were exceptionally high, it was easier to obtain ground turkey and turkey parts.
6. Marinating tougher cuts of meat in vinegar, soy sauce, or wine, and cooking with tomatoes helps to make the meat more tender.

Planked Ground Beef
1 lb. ground beef (454 g)
¼ cup oatmeal (25 g)
2 tablespoons wheat germ
¾ teaspoon salt
pinch of pepper
3 tablespoons nonfat dry milk
¼ cup water (59 ml)
½ teaspoon Worcestershire sauce (2.5 ml)

½ cup cheese, grated (46 g)
1 can mushrooms, sliced
1 onion, sliced
1 tomato, sliced

Combine the beef, oatmeal, wheat germ, salt, pepper, nonfat dry milk, water, and Worcestershire sauce, and form into patties. Between two patties place grated cheese, mushrooms, onion, and tomato slices. Broil patties on each side on a plank, platter, or cookie sheet. Then surround the patties with cooked green beans or lima beans and mashed potatoes or rice pilaff. You can make the pilaff or potatoes from a box or from scratch. Garnish with cherry tomatoes, green pepper slices, or black olives. This is an elegant looking meal. Serves 4.

Zucchini Boats

6 zucchini, parboiled, scooped from the shells and chopped
¾ pound ground beef, browned (340 g)
½ carrot, diced fine
½ cup celery, diced fine (80 g)
½ cup onion, diced fine (80 g)
1 egg, beaten
1 tablespoon wheat germ, bread crumbs, or cooked rice
½ teaspoon garlic salt
salt and pepper to taste
½ cup tomato juice (118 ml)
2 tablespoons Parmesan cheese, grated

Sauté the vegetables in 2 tablespoons (30 ml) oil. Pour off any leftover oil. Mix the meat, vegetables, wheat germ, egg, and seasonings. Fill the zucchini shells. Pour the tomato juice over the top and then sprinkle the tops with Parmesan cheese. Bake at 375° F (190° C) for 30 minutes or until the zucchini are just tender. Serves 4.

Barbecued Beef Roll-Ups

7–bone chuck roast with large section cut into long strips ½ in. (12 cm) thick
3 tablespoons parsley, chopped
2 green onions, chopped
½ teaspoon salt
dash pepper
1 teaspoon dry mustard

Sprinkle meat with a mixture of parsley, onions, salt, pepper, and dry mustard. Then roll up and secure with skewer. Cut about ½ in. (1.2 cm) thick for each roll-up. Broil in the oven or on outdoor barbecue or hibachi. Baste both sides with barbecue sauce.

Barbecue Sauce

1 can tomato sauce (8 oz/217 g)
1 teaspoon dry mustard
1 tablespoon vinegar (15 ml)
2 teaspoons Worcestershire sauce (10 ml)
¼ cup onion, minced (40 g)
1 clove garlic, minced
¼ cup ketchup (59 ml)
dash Tabasco sauce
¼ teaspoon salt
½ teaspoon paprika
½ teaspoon black pepper
2 cups water (470 ml)
½ cup butter or margarine (optional) (115 g)

Mix all ingredients and simmer for 30 minutes. Sauce should thicken a little.

Chuck Chile

7–bone chuck roast—pieces remaining after cutting for barbecued roll-ups

1 can red chile sauce or enchilada sauce (10 oz/280 g)
1 bouillon cube
2 cups water (470 ml)
1 to 2 tablespoons flour made into paste with a little cold water (optional)
½ onion, chopped

Cut the chuck into small square pieces. Then simmer beef with chile sauce, bouillon, and water for about 45 minutes or until tender. We prefer our chile unthickened, but you may like to add the flour paste. Serves 4–6.

Serve over rice, wrapped in flour tortillas, or as a filling for pocket bread.

Flank Steak (Filipino Style)

1 flank steak (medium)
4 tablespoons soy sauce or ½ teaspoon salt or to taste (60 ml)
dash of pepper
2 tablespoons vinegar or lemon juice (30 ml)
2 bay leaves, crushed
2 tablespoons cooking sherry (optional) (30 ml)

Marinate for 15 to 20 minutes.

3 cloves garlic, chopped
1 onion, chopped
1 small can tomato sauce with half amount of water (227 g)
24 olives, green with pits

Sauté garlic and onion until transparent, then add marinated meat. Cook until there is little juice left. Add tomato sauce and water. Let boil for 10 minutes then add olives and simmer until meat is done. Serves 4.

Pork Chop and Cabbage Casserole

3 tablespoons flour
1 large can evaporated milk (382 ml)
1 teaspoon sugar
½ large head of cabbage, sliced
4 pork chops, browned

Butter a large rectangular baking dish. Blend the flour with some of the milk, then add to the remainder of the milk. Stir in the sugar. Place the cabbage in the baking dish. Pour the milk mixture over the cabbage. Lay the pork chops on top. Bake at 350° F (177° C) for 1 hour. Serves 4.

Teriyaki

Cut any kind of meat into thin strips and place on wooden skewers. Marinate in the following sauce about 15 minutes before broiling. It is best to broil at table while eating. Serve with rice and French style green beans.

Sauce

1 cup soy sauce (235 ml)
⅓ cup water (78 ml)
⅓ cup sugar (62 g)
1 oz Japanese wine or sherry (optional) (30 ml)
2 teaspoons fresh ginger root, grated, or ¼ teaspoon ground
dash of monosodium glutamate
1 to 2 green onions, chopped

You can make this sauce ahead of time. Any leftover sauce can be frozen.

Pork with Peas and Bell Pepper

  1 teaspoon salt
½ cup water (118 ml)
  1 lb. pork steak, no bone (454 g)
  1 clove garlic
salt to taste
  1 box frozen peas
  2 ripe bell peppers (red), sliced medium thick

Cut meat into small pieces. Put in pot with water and salt. Boil until water evaporates. Do not simmer. Add garlic, keep stirring for two minutes, then add vegetables. Salt to taste.

Sweet and Sour Pork

  1 lb. pork, cut in cubes (454 g)
½ teaspoon soy sauce (2.5 ml)
  1 bell pepper, cut in small pieces
  1 large onion, sliced
  2 stalks celery, sliced
  4 teaspoons cornstarch
  2 or 3 slices pineapple, cut in chunks
  2 tablespoons sugar
  2 tablespoons vinegar (30 ml)
½ cup pineapple juice (118 ml)
½ teaspoon salt, or to taste
¼ teaspoon monosodium glutamate (optional)
¼ cup water (59 ml)
  3 tablespoons ketchup (45 ml)

Mix the pork with the soy sauce and cornstarch and spread on cookie sheet and brown in the oven at 350° F (177° C) for 20 minutes or until brown. Heat a little oil in a frying pan and sauté bell pepper, onion and celery for a few minutes. Then add pineapple, and take vegetables out and set aside.

Mix a little cornstarch, sugar, vinegar, soy sauce, pineapple juice, salt, monosodium glutamate, and water. Pour into the frying pan and cook until thickened. Add meat and vegetables and cook five minutes. Add three tablespoons of ketchup for coloring. Serves 4.

Pork Paprika (Filipino Style)

    1 lb. pork steak, ½″ (1.2 cm) thick, sliced in pieces 3 × 4 inches (7.5 × 10 cm) (454 g)
    2 teaspoons of salt
    3 bay leaves
4–6 cloves garlic, crushed
    4 tablespoons of vinegar (60 ml)
   ¼ teaspoon peppercorn, crushed
    2 teaspoons paprika

Mix all of the above together and marinate for 20 minutes.

1 cup water (235 ml)
1 medium sized onion, cut in 8 pieces
1 large sized potato, cut in 8 pieces

Mix these three ingredients; add to marinated meat and boil for 5 minutes; then simmer until meat is tender. Serves 4.

Chinese Style Flank Steak

1 flank steak (sliced very thin) (450 g to 675 g)
4 tablespoons soy sauce (45 ml) (60 ml)
2 teaspoons cornstarch or 4 teaspoons flour
⅛ teaspoon pepper
4 medium green peppers, cut in strips
4 tablespoons oil (45 ml)
1 teaspoon salt
¼ teaspoon monosodium glutamate (optional)

Mix beef with 2 tablespoons (30 ml) soy sauce, cornstarch, and pepper. Fry green peppers in 3 tablespoons (45 ml) oil. Remove, add remaining oil and fry beef until redness disappears. Add peppers, salt, monosodium glutamate, and remaining soy sauce. Serves 4–6.

Serve with rice.

Flank Steak Tacos

2 pounds flank steak (908 g)
½ onion, chopped
3 garlic cloves, minced or crushed
1 cup stock from meat (235 ml)
1 teaspoon salt
¼ teaspoon pepper
¼ teaspoon cumin
1 small can of tomato sauce (8 oz/227 g)

Place meat in skillet or pot with enough cold water to cover. Add onion and 2 cloves garlic and cook on low heat for 2-3 hours. Shred meat into bite-sized pieces and add salt, pepper, meat stock, cumin, 1 clove garlic, and tomato sauce. Cook until thickened; take out to serve. This will fill about 24 taco shells or several pocket breads or sopaipillas. Garnish with lettuce, tomatoes, etc. Serves 6–8.

Chicken Chop Suey

  1 chicken breast, cut up (336 g)
  2 tablespoons cooking sherry (30 ml) (optional)
1½ tablespoons soy sauce (22.5 ml)
dash of pepper

Marinate for ten minutes.

  1 teaspoon ginger, chopped fine
  1 clove garlic, sliced
  1 onion, chopped
 ½ bell pepper, sliced
  1 lb. bean sprouts (454 g)
10 medium mushrooms, sliced
  1 medium carrot, sliced
  2 stalks celery, cut up
  1 tomato, cut up
  3 chicken bouillon cubes
 ½ cup chicken broth (118 ml)
  2 tablespoons cornstarch
  1 teaspoon sugar

Heat 2 tablespoons of oil in frying pan over high flame, add ginger, garlic, and onions and cook until transparent. Add mushrooms, then chicken, and sauce. Cover and cook about two minutes. Add some broth and bouillon and sugar which have been mixed together. Add celery, carrots, bell pepper, ½ teaspoon monosodium glutamate, and cover about two minutes. Add bean sprouts, more broth, and stir. Mix 2 tablespoons cornstarch and 2 tablespoons water. Add to pan and stir until thickened. Season with soy sauce as desired. Add tomato wedges; mix well. Cook 1 more minute. Serve with rice. Serves 4.

Pineapple Chicken

 2 cups sliced raw chicken breasts or pieces of cooked chicken (672 g)
18 pineapple cubes (canned)
 1 green pepper, cut into 1-inch (2.5 cm) long thin slices
 1 cup celery, sliced (160 g)
salt and pepper to taste

Sauce

⅔ cup 50 grain white vinegar (155 ml)
 1 cup apricot nectar (235 ml)
 1 cup brown sugar (210 g)
 1 teaspoon Worcestershire sauce
 1 cup ketchup (235 ml)
 1 teaspoon cornstarch

Prepare sauce first by combining all ingredients except cornstarch. Simmer for 30 minutes. Thicken with cornstarch diluted in a little water. Sauté chicken in a little oil (do not brown), add all other ingredients and stir constantly. Add Sweet and Sour sauce, salt and pepper to taste. Serves 6.

Sesame Chicken

  2 tablespoons flour
 ½ teaspoon salt
 ¼ teaspoon pepper
  2 whole chicken breasts (you actually get 4 pieces) (672 g)
  2 eggs
 ¼ cup milk (59 ml)
  3 tablespoons sesame seeds + 4 tablespoons flour

Shake flour, salt, pepper, and the chicken breasts in a bag. Beat the eggs and milk. Dip each flour coated breast into egg mix; then roll in sesame mix and fry in hot oil. Serve with light supreme sauce.

Light Supreme Sauce

   3 tablespoons butter or margarine (45 g)
   2 tablespoons flour
 1½ cups chicken broth (353 ml)
salt and pepper to taste
   1 egg yolk

Melt the butter in a saucepan. Stir in the flour to make a smooth paste. Gradually add the chicken broth and cook and stir to make a smooth sauce. Season with salt and pepper. Beat the egg yolk. Gradually whisk about one-half cup of the hot sauce into egg yolk. Gradually beat yolk mixture back into sauce and cook over low heat, whisking, until sauce is thickened and smooth. Pour sauce over chicken. Serves 4.

Almond Chicken

- 2 whole chicken breasts (672 g)
- 3 tablespoons salad oil (45 ml)
- 1 cup celery, sliced (160 g)
- 1 clove garlic, minced
- 2 envelopes instant chicken broth or 2 chicken bouillon cubes
- 1½ cups water (353 ml)
- 1 tablespoon soy sauce (15 ml)
- 1 tablespoon chopped crystallized ginger or ¼ teaspoon ginger
- 1 package frozen peas
- 2 tablespoons flour
- ½ cup toasted slivered almonds (80 g)
- 3 cups hot cooked rice (615 g)

Pull skin from chicken breasts and bone and slice meat into long thin strips. Sauté chicken in oil, stirring constantly for 5 minutes. Stir in celery and garlic, sauté 3 minutes more. Then stir in bouillon, water, soy sauce, and ginger, and bring to boil. Add peas and simmer for 5 minutes. Smooth the flour in a little water and stir into the chicken mixture. Sprinkle with monosodium glutamate if desired. Turn onto serving plate over rice and garnish with almonds. Serves 6.

Chicken Casserole

⅔ cup stuffing mix (136 g)
1 package frozen green beans, slightly cooked and drained
4 tablespoons slivered almonds (40 g)
2 cups cooked chicken or turkey (454 g)
1½ cups stuffing mix (105 g)
½ cup water (118 ml)
2 tablespoons butter
1 can cream of mushroom soup and ½ can of water (301 g)

Place the first four ingredients in a casserole in layers; mix the stuffing mix with ½ cup water and spread over the meat and vegetables. Mix the mushroom soup with the half can of water and pour over the top. Bake in a 400° F (205° C) oven for 30 minutes. Serves 4.

Special Chicken Casserole

1 box wild rice mix
1 chicken, cooked and boned
1 can cream of celery soup (301 g)
½ cup water (118 ml)
1 small jar pimiento, chopped
1 small can water chestnuts, sliced thin (168 g)
½ cup salad dressing (118 ml)
1 large can string beans, French cut (454 g)

Prepare wild rice mix according to package directions

in a large pan and leave it in the pan. Mix in the other ingredients and place in a greased casserole and bake covered at 350° F (177° C) for 35 minutes. Serves 6.

This is a great main course for a special dinner. Serve bread sticks and a gelatin salad with it.

Variations: Use 1 cup brown or white rice in place of wild rice mix. Add 2 chicken bouillon cubes or 2 cups (474 ml) chicken broth and ¼ teaspoon parsley flakes. This would cost less to make.

Parslied Turkey Tenderloin

    2 tablespoons oil (30 ml)
    1 medium onion, sliced
    4 slices of turkey tenderloin or breast, cut about ½-inch (12 cm) thick
flour for rolling
    ¼ cup water (59 ml)
    1 tablespoon lemon juice (15 ml)
    1 tablespoon parsley, chopped
    1 teaspoon salt
    1 tablespoon Parmesan cheese, grated

Sauté the onion in the oil until tender. Then remove the onion from the skillet. Add more oil if there is not enough left to coat the bottom of the skillet. Coat the pieces of turkey with flour. Brown, turning once. Add onion, water, lemon juice, parsley, and salt. Cover and simmer for about 15 minutes. The turkey should be tender. Sprinkle with Parmesan cheese. Serves 4.

Cassoulet

2 cups cubed left-over meat or luncheon meat (454 g)
salt and pepper
3 tablespoons vegetable oil (45 ml)
1 medium onion, chopped
1 to 1½ cups boiling water (235 ml)
1 garlic clove
1 herb bouquet (assorted herbs tied in cheese cloth or in tea ball)
2 cans vegetarian beans in tomato sauce (1 lb/454 g each)
½ lb. sausage link, browned (228 g)
½ cup chile sauce (118 ml)
3 tablespoons buttered bread crumbs

Sauté meat in hot oil until browned. Drain off excess oil and add chopped onion. Cook onion until golden brown. Cover meat with boiling water; add garlic, salt and pepper, and herb bouquet. Simmer 15 minutes. Remove herbs and garlic. Rub inside of casserole with garlic. Place layer of beans in casserole; then a layer of browned meat and sausage. Cover with layer of baked beans. Add chili sauce to liquid in which meat was cooked; pour over beans and meat. Top with buttered crumbs. Bake at 350° F (177° C) for 45 minutes. Serves 6–8.

Peppers Parmigiana

4 large green peppers
1 can corned beef hash (1 lb/454 g)
¼ cup grated American cheese (23 g)
¼ cup ketchup (59 ml)
1 clove garlic, minced
1 teaspoon oregano or Italian seasoning
2 teaspoons Parmesan cheese, grated

Wash peppers; cut off stem ends; scoop out cores and seeds. Parboil in salted water for 5 minutes; drain. Combine corned beef hash, grated American cheese, ketchup, garlic, and oregano. Fill peppers. Sprinkle with grated Parmesan cheese. Bake at 375° F (190° C) for 30 minutes. Serves 4.

Old Fashioned Vegetable Soup
2 pounds soup bone and boiling beef (908 g)
4 quarts water (3760 ml or 3.8 l)
1 medium onion cut in 7 rings
1 tablespoon beef bouillon
2 tablespoons ketchup
1 can stewed tomatoes (454 g)
⅓ cup lentils (60 g)

Combine the ingredients above in a large pot with tight lid. Bring to a hard boil and let boil for 5 minutes. Cover tightly and let simmer for 3 hours. Then bring soup to a hard boil and add:
1 carrot, diced
1 stalk celery, diced
½ cup cut string beans, fresh or frozen (70 g)
1 cup dry egg noodles (60 g)
¼ cup rice (50 g)
1 medium potato, cut up
1 small rutabaga or turnip, cut up
1 cup cabbage, cut up or shredded (70 g)
½ cup peas, fresh or frozen (70 g)
8 pods of okra, cut up
1 teaspoon salt or to taste
pinch of garlic salt (optional)

Boil for 2 minutes, then cover pot tightly and simmer for 45 minutes. Serve in bowls and top with grated Parmesan cheese. Serves 8.

Serve with French bread or garlic bread and pudding. Note: This recipe is an original of my mother's. Whenever she prepares it for family or friends, she always receives smash reviews.

Turkey Scallopini

2½ lbs. turkey tenderloin (1135 g or 1.1 kg)
¼ cup Parmesan cheese, grated (25 g)
½ cup flour (60 g)
2 teaspoons salt
½ onion, chopped fine
½ cup oil (118 ml)
½ cup red wine (118 ml)
8 medium mushrooms, sliced
1 can tomatoes (1 lb/454 g)
1 teaspoon sugar
¼ teaspoon thyme

Cut turkey tenderloin into 1 in. (2.5 cm) cubes. Roll in cheese and beat cheese in. Roll in flour to which one-half teaspoon of salt and one-eighth teaspoon of pepper have been added. Sauté onion in oil until tender; sauté turkey in same oil until golden brown. Pour in wine; turn off heat and cover immediately and let stand 2 minutes. Add remaining ingredients; cover again and simmer for 1 hour or until tender. Serves 6.

Turkey Corn Chowder

½ large onion, chopped
3 tablespoons oil (45 ml)
½ lb. ground turkey (230 g)
2 cakes tofu, well drained (½ lb/230 g)
1 can corn (12 oz/336 g)
1 large potato, diced
1 can tomatoes (1 lb/454 g)
1½ teaspoons salt
½ teaspoon pepper
2 teaspoons sugar
3 cups water, boiling (707 ml)
⅔ cup canned evaporated milk (78 ml)

Sauté onion in oil in large kettle until transparent. Push to one side and add ground turkey which has been mixed with the tofu, and brown. Add remaining ingredients except canned milk. Stir well and cover. Bring just to a boil and reduce heat to simmer until potatoes are tender, about 20 to 30 minutes. Remove from heat and slowly stir in canned milk. Serves 6.

Stuffed Chicken Thighs

6 chicken thighs, boned
salt to taste
1 recipe Cornbread Stuffing (see GRAINS)
peanut oil for basting

Cut tendons at each end of the thigh bone. Take thumb and push meat off the bone. Fill the cavity with stuffing. Sprinkle the outside with salt and baste with peanut oil. Place in a shallow baking pan and bake at 350° F (177° C) for about 1 hour. Serves 6.

Serve with cranberry sauce. Leftover stuffing can be wrapped in aluminum foil and baked with the chicken. This would be good roasted in a covered barbecue grill.

Fish

GENERAL INFORMATION

In many households fish is not a very popular food. One reason is that many people do not prepare it correctly. Another may be that it is not really fresh when they purchase it and cook it. When it is fresh and prepared correctly, fish is quite delicious.

In addition to the nutritive contribution made by fish, another advantage in incorporating it into our meals is that many fish and seafoods are moderately priced and a relatively inexpensive source of protein.

There is a wide range in the price of fish and other seafoods. Salmon, halibut, sole, lobster, and crab are some of the more expensive, whereas mackerel, turbot, bonita, tuna, butterfish, and clams are some of the less expensive.

Canned mackerel is about the cheapest fish I have found. I first started using it after a student asked me if it was fit for human consumption. She couldn't understand why it was so cheap. Consequently, I began to experiment with recipes for canned mackerel. It has a greater oil content than most fish and tends to have a little stronger flavor. It mixes well with an unflavored textured vegetable protein.

If you live near relatively unpolluted waters and someone in your family enjoys fishing, you can stretch your budget by catching your own fish. There are few other types of recreation which give you something other than pleasure for your dollars spent.

NUTRITIVE VALUE

Fish contains a high percentage of good quality protein, and can help to improve the protein value of vegetable proteins. An average 3 oz. (84 g) serving provides about one-third of an adult's daily protein requirement. Fish also provides important amounts of iron, copper, B vitamins, and iodine. Salt water fish is one of our few dependable food sources of iodine other than iodized salt. If the bones are eaten, canned fish such as sardines, salmon, and mackerel provide valuable amounts of calcium.

Another advantage of using fish is that it is relatively low in calories and saturated fats. Even the fatty fish are only about 10 percent fat, and this is relatively polyunsaturated. Therefore, fish is good for use in calorie controlled and fat controlled diets.

In the not too distant future a form of fish known as fish flour will probably be available for use in most kitchens. It will be used mainly to boost the protein value of foods. Since it is tasteless it can be added to almost anything, particularily to grain and legume dishes. Fish flour will also make a great contribution to the iron and calcium intake since the whole fish is used in producing the flour. I regret that I do not have any flour available at this time to include in recipes.

COOKING TIPS

1. Do not overcook. Fish is done when it flakes easily with a fork.
2. If you catch your own fish and wish to freeze it, freeze it in a container filled with water. Milk cartons make good containers.
3. If the bones are a drawback to your family's acceptance of fish, fillet it. Special tools can be purchased but a good sharp knife is all that is really necessary.

Table XI
Nutrient Content of Selected Fish, 100 g [1]

Food	Calories	Protein gm	Carbohydrate gm	Fat gm	Calcium mg	Iron mg	Vitamin A I.U.	Thiamin mg	Riboflavin mg	Niacin mg	Vitamin C mg
Cod, cooked	170	28.5	0	5.3	31	1.6	180	.08	.11	3.0	2
Mackerel, cooked [2]	236	21.8	0	15.8	6	1.2	430	.15	.27	7.6	—
Mackerel, canned [2]	183	19.3	0	11.1	185	2.2	430	.06	.21	5.8	—
Salmon, cooked	182	27.0	0	7.4	—	1.2	160	.16	.06	9.8	—
Salmon, canned	141	20.5	0	5.9	196	.8	70	.03	.18	8.0	—
Tuna, canned (oil)	197	28.8	0	8.2	8	1.9	80	.05	.12	11.9	—
Fish Flour	336	78.0	0	.3	4610	41.0	—	.07	.62	2.2	—

[1] Values taken from *Composition of Foods*, Agriculture Handbook No. 8, Agricultural Research Service, United States Department of Agriculture, 1963.
[2] Atlantic mackerel.

To Poach Fish

(For use in fish salads or other recipes)

1 lb. fish fillets (454 g)
1 onion
1 stalk celery, including tops
1 carrot
1 teaspoon salt
water, just to cover

Put all ingredients in a pan. Bring to boil. Reduce heat, cover, and simmer until fish is just tender. Remove from heat, drain, and remove fish for use as desired. Makes 4 servings.

Poached White Fish

1½ lbs. cod, flounder, haddock or other white fish fillets (681 g)
⅓ cup nonfat dry milk (28 g)
1 cup seasoned chicken bouillon, clam juice, or white wine (235 ml)
mushroom sauce (below)

Wipe fish fillets with damp cloth. Arrange in four portions in lightly greased skillet. Combine remaining ingredients in a saucepan and bring to a boil. Pour over fish fillets and simmer, covered, until fish is tender, about 15 minutes. Serve with Mushroom Sauce. Serves 4.

Mushroom Sauce

1 cup skim milk or 1 cup liquid used in poaching fish (235 ml)
3 tablespoons finely minced onion
1 tablespoon finely chopped parsley
1 teaspoon dehydrated grated lemon rind
½ teaspoon salt
dash of pepper
dash of savory
1 can sliced mushrooms or 6 fresh mushrooms, sliced (56 g)
2 tablespoons cornstarch or 3 tablespoons flour
2 tablespoons sherry (optional) (30 ml)

Combine liquid, onion, parsley, seasonings, and drained mushrooms. Blend together and stir in mushroom broth and cornstarch. Cook, stirring constantly, until thickened. Add sherry and serve over fish.

Chesapeake Bay Fish Salad

  1 lb. fish fillets, cooked (454 g)
  1 tablespoon onion, finely chopped
  2 tablespoons green pepper, finely chopped
¼ cup salad oil (59 ml)
  2 tablespoons wine vinegar (30 ml)
⅛ teaspoon garlic powder
⅛ teaspoon oregano leaves
  1 teaspoon parsley flakes
salt and pepper to taste
  2 tablespoons sour cream

Cut fish into very small pieces. Put into a bowl and add onion and green pepper. Add oil, vinegar, garlic powder, oregano, and parsley. Mix well. Add salt and pepper. Refrigerate until thoroughly chilled. Just before serving, stir in sour cream. Makes about 2 cups (474 ml) of salad.

Tuna and Alfalfa Sprouts Salad

  1 cup alfalfa sprouts (40 g)
  1 can tuna or 1 cup poached fish (182 g)
¼ cup mayonnaise or salad dressing (60 g)
  1 teaspoon lemon juice, optional (5 ml)
salt and pepper to taste

Mix ingredients thoroughly. Serve on bread or crackers, or stuff a tomato.

Variation: Add chopped parsley, chopped onions, and/or chopped pickles.

Fish Stew

1 pound firm fish fillets (454 g)
1 potato, sliced
1 tomato, sliced
1 green pepper, chopped
1 stalk celery, chopped
½ medium onion, chopped

Combine all ingredients in large pot and cover with water. Bring to a boil; then reduce heat and simmer until fish flakes and vegetables are tender. Serves 4.

Serve with bread or rolls.

Variation: Add sliced carrots or any other vegetables you like in soup or stew.

Clam Chowder

4 tablespoons butter or margarine (60 g)
4 tablespoons flour (40 g)
1 onion, chopped
2 potatoes, diced
2 carrots, diced
4 cups hot water (940 ml)
1 can of clams (8 oz/227 g)
salt and pepper to taste

Melt butter or margarine and add enough flour to make a thick white sauce. Brown it a little. Add chopped onions and diced potatoes, mix with butter and flour. Add the carrots to the potato and onion mixture. Add hot water, as much as needed to create the consistency you like your soup. Boil. Reduce heat to medium and cook for 20 to 30 minutes, until potatoes are tender.

Just before serving add clams and salt and pepper to taste. Boil and serve. (Clams will become tough if added too early.) Serves 4.

This can be a meal in itself. Chowder tastes great with French bread or biscuits.

Filet of Fish Grenobloise

 1 lb. fish fillets, salted (454 g)
 1 egg, beaten with 1 tablespoon of water or milk
¼ cup flour (62 g)
 2 tablespoons butter or margarine

Dredge fish in flour, dip in egg, and pan fry in clarified butter. Sprinkle with lemon juice. Serves 3–4.

Halibut Steak with Sour Cream Sauce

 2 halibut steaks (¾ in/1.9 cm thick, 1–1½ lbs/681 g)
salt and pepper
⅓ cup sour cream (76 g)
 2 tablespoons dry bread crumbs
⅛ teaspoon garlic salt
 1 tablespoon chives or green onion tops, chopped
 2 tablespoons Parmesan cheese, grated
¼ teaspoon paprika

Place halibut steaks close together in a shallow buttered baking dish. Sprinkle with salt and pepper. In a bowl, mix together the sour cream, bread crumbs, garlic salt, and chives or onions. Spread mixture over fish. Sprinkle with cheese, then paprika. Bake uncovered in a 400° F (205° C) oven for about 20 minutes or until fish flakes with a fork. Garnish with parsley and lemon wedges if you wish. Serves 2–3.

Spanish Style Baked Fish

 1 lb. frozen or fresh fish fillets (454 g)
 ½ onion, chopped
 ⅓ cup celery, chopped (50 g)
 2 tablespoons margarine or butter
 1 small can tomato sauce (227 g)
 1 tablespoon all-purpose flour
 ½ teaspoon paprika
 ¼ teaspoon salt
 1 tablespoon lemon juice (15 ml)
 ½ teaspoon chili powder

Thaw fish if frozen. In saucepan cook onion and celery in margarine until tender but not brown. Stir in remaining ingredients *except* fish. Simmer uncovered for 10 minutes. Cut fish crosswise into four portions; arrange in shallow baking pan. Spoon sauce over the fish. Bake uncovered at 350° F (177° C) for 30 minutes. Serves 4.

Shrimp Chow Mein

   2 stalks celery, sliced thin
   5 or more mushrooms, sliced
   3 green onions, chopped
 ½ pound prawns, diced small, or fresh shrimp (227 g)
   1 pound pork butt, sliced in small strips (454 g) (optional)
   1 cup chicken broth or bouillon (235 ml)
   2 tablespoons soy sauce (30 ml)
salt to taste
dash of pepper
pinch of sugar
1½ tablespoons cornstarch or 3 tablespoons flour made into paste with a little water
dash of monosodium glutamate—optional
oil for sauteing and stir frying

Stir fry the celery, onion, mushrooms, salt, and pepper until cooked but still firm. Then remove and set aside. Add more oil to pan and stir fry pork slices until brown. Season with salt and pepper. Move the pork to the side of the pan and add shrimp. Stir fry briefly. Then add vegetable mixture, chicken broth, soy sauce, sugar, monosodium glutamate, and cornstarch paste, and bring to boil. Serves 4–6.

To serve, mix in cooked chow mein noodles or pour over crispy fried chow mein noodles.

Charcoal Grilled Fish Steaks

1 lb. (454 g) lingcod, rock cod, red snapper, ocean perch or other similarly firm-fleshed fish cut into fillets one-half to three-quarters inch thick.

Marinate fillets for one-half hour in:

½ cup dry white or red wine (118 ml) (optional)
1 teaspoon fine herbs
1 large clove garlic, crushed
salt and pepper to taste

Place fish on well greased wire grill. Brush fish with olive oil. Cook about 4 inches (10 cm) from moderately hot coals for 4 minutes, basting with marinade. Turn fish and cook for 4 minutes, basting with marinade. Turn fish and cook for 4 minutes, again brushing with olive oil and marinade. Fish will flake easily with fork when done. Serves 2–3.

If desired, reduce marinade by about one-half by boiling and serve on fish.

Sweet and Sour Broiled Fish

½ cup lemon juice (118 ml)
¼ cup oil (59 ml)
½ teaspoon pepper
2 tablespoons onion, minced
1 teaspoon dry mustard
2 tablespoons brown sugar
1 lb. fish fillets (454 g)

Mix all ingredients except fish together until sugar is dissolved. Placing fish in shallow pan about 2 inches from heat, broil fish on both sides until brown and tender, basting frequently with sauce. Serves 3–4.

Sesame Fish

2 lb. fish fillets (908 g)
salt and pepper
1 egg
2 tablespoons undiluted evaporated milk (30 ml)
½ cup flour (62 g)
½ cup dry bread crumbs (25 g)
4 tablespoons sesame seeds (40 g)
⅓ cup oil (78 ml)
⅓ cup fresh parsley, chopped (15 g)
¼ cup lemon juice (59 ml)
½ small onion, chopped fine

Season fish with salt and pepper. Beat egg and milk together. Dip each fillet lightly into flour, then into liquid, then into bread crumbs and seeds mixed together. Heat oil until hot, reduce heat, and fry 3 to 8 minutes, depending on thickness of fish. Absorb excess oil by draining on paper towels. Combine parsley, lemon juice, and onion, and pour over fish before serving. Serves 5–6.

Fish-Vegetable Bake

  2 lbs. fish fillets, fresh or frozen (908 g)
  2 teaspoons salt
  ¼ teaspoon pepper
  1 can whole potatoes, drained (1 lb/454 g)
  1 can whole onions, drained (1 lb/454 g)
  1 package frozen mixed vegetables (280 g)
  2 tablespoons lemon juice (30 ml)
  1 can tomato soup (10¾ oz/305 g)

Thaw frozen fish. Cut into 6 serving-size portions. Cut six pieces of heavy duty aluminum foil, 12 × 12″ (30 × 30 cm) each. Grease lightly. Place one piece of fish on each piece of foil. Season with salt and pepper. Divide remaining ingredients equally among the packages of fish, using the soup last. Bring the foil up over the fish and seal the edges by making double folds in the foil to hold the juices. Place the packages inside a covered barbecue or in the oven. Cook in a slow oven, 300° F (149° C) for 15 minutes; open packages by cutting a crisscross in the top of each package and fold the foil back. Continue to cook for 10–15 minutes longer until the fish flakes easily when tested with a fork. Serves 6.

Smoked Fish with Brown Rice and Mushroom Stuffing

1 dressed snapper, or other large fish (3 to 4 lb/1362 g to 1816 g)
2 teaspoons salt
¼ teaspoon pepper
4 slices bacon, precooked
2 green onions with tops, thinly sliced
Brown rice and mushroom stuffing

Thaw frozen fish. Clean, wash, and dry fish. Sprinkle inside and out with salt and pepper. Stuff fish loosely. Close opening with small skewers or toothpicks. Place precooked bacon on top of fish and sprinkle with sliced onions. Place fish on well-greased grill inside a covered barbecue or in the oven. Cook in a slow oven 325° F (163° C) for approximately 1 hour or until fish is done and flakes easily when tested with a fork. Remove skewers. Serves 6.

Brown Rice and Mushroom Stuffing

1 cup brown rice or 1 package brown and wild rice mix (200 g)
4 tablespoons margarine or cooking oil (60 g)
½ onion, chopped (medium size)
2 stalks celery, chopped fine (medium size)
1 small can mushrooms, sliced (56 g)
1 egg (optional)
¼ cup parsley, chopped (10 g)
¼ teaspoon pepper

Cook brown rice according to directions on package, using chicken or fish stock instead of water. Sauté vegetables in margarine or cooking oil until vegetables are tender. Combine all ingredients and mix thoroughly. Makes approximately 2½ cups stuffing.

Mackerel Crescent Swirls

  1 can crescent refrigerator rolls
  1 can mackerel (420 g)
  ½ can tomato sauce (113 g)
  ¼ medium onion, chopped
  ½ cup Monterey Jack or Cheddar cheese, grated (46 g)

Flatten crescent rolls into one long, rectangular sheet. Mix mackerel, tomato sauce, and onion. Spread the mixture over the rolls. Sprinkle the cheese over the top. Then roll up into one long roll. Cut into one-inch (2.5 cm) pieces and place on greased baking sheet. Bake at 375° F (190° C) for about 20 minutes. Serves 4–6.

Variation: Use your own biscuit recipe rolled very thin or use packaged biscuits. You may also spread mixture on English muffins. Mix the mackerel with ¼ cup (20 g) rehydrated unflavored TVP to serve more people.

Fish in Shrimp Soup

  1 can condensed shrimp soup (305 g)
  ¼ cup dry white wine (59 ml) (optional)
  ¼ cup lemon juice (59 ml)
  1½ lbs. butterfish or other inexpensive fish such as cod or snapper (671 g)

Place fish in baking dish. Mix other ingredients and pour over fish. Bake in moderate oven (350° F) (177° C) for 40 minutes, basting occasionally. Serves 6.

French Fried Fish Fillets

peanut oil for deep fat frying
1 lb. or more fish fillets like perch, sole, flounder or bass (454 g)
salt
1 egg
1 tablespoon milk (15 ml)
cornflake crumbs or cracker crumbs

Salt the fish. Beat the egg and milk together. Dip the fish in egg mixture and then roll in crumbs. Fry in hot fat. Drain on paper towels. Serves 4.

Southern-Style Fried Catfish

4 catfish, skinned, cleaned, and heads removed
salt to taste
½ cup cornmeal (67 g)
brown paper bag
polyunsaturated oil for deep fat frying

Salt the fish. Put corn meal in paper bag. Drop in the fish and shake. Take out the fish and fry in hot oil. I find it most successful to drain the fish on cake cooling pans placed over a cookie sheet. The fish can be kept warm in oven that way. Serves 4.

References

Amino Acid Content of Foods and Biological Data on Proteins, Food and Agricultural Organization of the United Nations, Rome, 1970.

Assorted Literature, California Dry Bean Advisory Board, Dinuba, California.

Assorted Literature, California Turkey Industry, Modesto, California.

Assorted Literature, Worthington Foods, Inc., Worthington, Ohio.

Assorted Materials, Consumers Cooperative of Berkeley, Inc., Berkeley, California.

Assorted Materials, Kansas Wheat Commission, Hutchinson, Kansas.

Assorted Materials, Pacific Gas and Electric Company, California, 1973.

Assorted Materials, Seafood Marketing Authority, Annapolis, Maryland, 1973.

Bogert, L. Jean, Briggs, George M., and Calloway, Doris Howes, *Nutrition and Physical Fitness*, W. B. Saunders, Philadelphia, Pa., 1973.

Casella, Dolores, *A World Of Breads*, David White Company, New York, 1966.

Deutsch, Ronald M., *The Family Guide to Better Food and Better Health*, Meredith Corporation, Des Moines, Iowa, 1971.

Hamdy, M. M., "Nutritional Aspects in Textured Soy Proteins," *Journal of American Oil Chemists' Society, 51*: 85A, January 1974.

Kellar, Richard L., "Defatted Soy Flour and Grits," *Journal of American Oil Chemists' Society, 51*: 77A, January 1974.

Lappe, Frances Moore, *Diet for a Small Planet*, Ballantine Books, Inc., New York, 1971.

Mattel, Karl F., "Composition: Nutritional and Functional Properties and Quality Criteria of Soy Protein Concentrates and Soy Protein Isolates," *Journal of American Oil Chemists' Society, 51*: 81A, January 1974.

Orr, M. L., and Watt., B. K., *Amino Acid Content of Foods*, Home Economics Research Report No. 4, U. S. Department of Agriculture, 1957.

Private Communication, Corning Company Representative, American Home Economics Association Convention, Los Angeles, California, 1974.

Products Fact Sheet, The Quong Hop Company, South San Francisco, California, 1973.

Protein-ettes (pamphlet), The Creamette Company, Minneapolis, Minn., 1974.

Recommended Dietary Allowances, 1973 Revision, Food and Nutrition Board, National Research Council, National Academy of Sciences, Washington, D.C.

San Jose Peninsula Dietetics Association, *Proceedings: Meeting*, Los Altos, California, February 1974.

Stare, Frederick J., and McWilliams, Margaret, *Living Nutrition*, John Wiley & Sons, Inc., New York, 1973.

U. S. Department of Agriculture, *Composition of Foods*, Agriculture Handbook No. 8, 1963.

U. S. Department of Agriculture, *Nutritive Value of Foods*, Home and Garden Bulletin No. 72, 1971.

U. S. Department of Agriculture, *Soybeans in Family Meals*, Home and Garden Bulletin No. 208, 1974.

U. S. Department of Agriculture, *Vegetables in Family Meals*, Home and Garden Bulletin No. 105, 1971.

University of California Extension, *Proceedings: Conference on Consumer Concerns in Food Processing*, Hayward, California, May 1974.

Index

Almond chicken, 145

Banana nut bread with wheat germ, 109
Banana pudding or pie filling, 124
Barbecued beef roll-ups, 136
Beans
  Dried beans and dried peas (any kind), 77
  Falafel, 76
  Fava with tomato sauce, 77
  Garbanzo beans (chick peas) and chicken, 80
  Pinto beans, 78
  Pinto beans, Portuguese style, 78
  Refried beans, 78
  Three bean casserole, 79
  Three bean casserole with cheese, 79
Beef
  Barbecued beef roll-ups, 136
  Beef, ground, planked, 134
  Beef or pork, with tofu, 46
  Beef pie, ground, with TVP, 65
  Brown rice and beef, 90
  Cabbage rolls, stuffed with tofu, 51
  Calypso skillet with TVP, 59
  Cannelloni with TVP, 66
  Cassoulet, 148
  Chipped beef casserole, 92
  Chuck, chili, 136
  Enchiladas with TVP, 60
  Hamburgers with tofu, 44
  Lasagne with tofu, 49
  Manicotti stuffed with tofu, 48
  Meat and potato roll-ups with TVP, 64
  Meatballs with tofu, 45
  Mexican casserole with TVP, 62
  Peppers stuffed with corned beef hash, 148
  Steak, flank, Chinese style, 141
  Steak, flank, Filipino style, 137
  Steak, flank, tacos, 141
  Tacos with TVP, 63
  Teriyaki, 138
  Tournament supreme with TVP, 66
  Zucchini boats, 134
Biscuits, 102
Biscuit bake, Parmesan, 102
Biscuit mix, 101
Biscuit mix, higher protein, 102
Blackeye peas and rice, 80
Breads, 96–115
  Cooking tips, 97
  General information, 96
  Nutritive value, 96
Brown bread, quick, 107
Brownies, carrot, 113
Brown rice and beef, 90

170 INDEX

Bulgar pilaff, 86
Bulgar, preparation of, 86
Bulgar-Tabouli, 91
Butter top coffee cake, 104

Cabbage rolls stuffed with tofu, 51
Calypso skillet with TVP, 59
Cannelloni with TVP, 66
Cashew nut casserole, 95
Cassoulet, 148
Cheese and Eggs, 116–131
Cheese bread, quick, 103
Cheese chops, 128
Cheese souffle, 126
Cheese sourdough French bread, 114
Chicken
  Almond, 145
  Casserole, 146
  Casserole, special, 146
  Chop suey, 142
  Curry with TVP, 56
  Garbanzo beans (chick peas), 80
  Pineapple, 143
  Sesame seeds, 144
  Thighs, stuffed, 151
Chilaquilas, 121
Chile Rellenos, 120
Chile soybeans, 33
Chipped beef casserole, 92
Chop suey, chicken, 142
Chow mein, shrimp, 160
Chuck, chili, 136
Clam chowder, 158
Clam Marinara sauce, 61
Coffee cake, butter top, 104
Coffee cake muffins, Sunday morning, 100
Coffee cake with tofu, 53
Cornbread biscuit stuffing, 100
Cornbread, East Texas, 105
Corn Quiche Lorraine, 126
Croquettes, fish, 44

Cucumber soup with tofu, 42

Egg Fu Yung with sauce, 124
Eggplant, Parmesan, 128
Eggplant stuffed with cheese, chive and egg, 128
Eggs and Cheese, 116–131
  Cooking tips, 118
  General information, 119
  Nutritive value, 116
Enchilada sauce, 60
Enchiladas with TVP, 60

Falafel, 76
Fava beans with tomato sauce, 77
Fettucine with cream, 93
Fish, 152–166
  charcoal grilled fish steaks, 161
  Chesapeake Bay fish salad, 157
  Clam chowder, 158
  Clam Marinara sauce with TVP, 61
  Cooking tips, 153
  Filet of fish Grenobloise, 159
  Filet, french fried, 166
  Fish croquettes with tofu, 44
  Fish in shrimp soup, 165
  Fish salad or sandwich spread with tofu, 50
  Fish stew, 158
  Fish-vegetable bake, 163
  General information, 153
  Halibut steak with sour cream sauce, 159
  Mackeral crescent swirls, 165
  Mackeral low cost fish cakes with TVP, 58
  Mackeral sandwich spread with tofu, 50
  Nutritive value, 153

INDEX

Sesame fish, 162
Shrimp chow mein, 160
Smoked fish with brown rice mushroom stuffing, 164
Southern style fried catfish, 166
Spanish style baked fish, 160
Sweet and sour broiled fish, 162
Tofu Sea Island salad, 52
To poach fish, 155
Tuna and alfalfa sprouts salad, 157
White fish, poached, 156
Flank steak, Chinese style, 141
Flank steak, Filipino style, 137
Flank steak tacos, 141
Fresh green soybeans in butter, 30

Gluten, preparation of, 87
Gluten steaks, chicken fried, 88
Gluten won tons, 88
Gold and green casserole, 129
Golden delight salad with soybeans, 30
Grains, Nuts, and Seeds, 82–95
  Cooking tips, 84
  General information, 82
  Gluten, preparation of, 87
  Nutritive value, 82
Graham cracker crust, 119
Granola, 89
Granola cookies, 90
Green chile and cheese casserole, 123
Green pepper stuffed with lentils, 72
Gumbo soup with soybeans, 32

Halibut steak with sour cream sauce, 159
Hamburgers with tofu, 44
High protein bread, 112
Hi-protein vegetable bread, 112
Huevos Rancheros (ranch style eggs), 122
Huevos Reuveltos (green chile omelet), 122

Lasagne with tofu, 49
Legumes, 68–81
  Blackeye peas and rice, 80
  Chicken and garbanzo beans (chick peas), 80
  Cooking tips, 69
  Dried beans and dried peas (any kind), 77
  Falafel, 76
  Fava beans with tomato sauce, 77
  General information, 68
  Green pepper stuffed with lentils, 72
  Lentils, basic, 70
  Lentils, loaf with tomato and cheese sauce, 72
  Lentils, puree, 71
  Lentils, spiced, baked, 73
  Lentils, stewed, 71
  Nutritive values, 69
  Pinto beans, 78
  Pinto beans, Portuguese style, 78
  Refried beans, 78
  Spareribs with lentils, 74
    to cook, 75
    to saute, 75
  Three bean casserole, 79
  Three bean casserole with cheese, 79
  Vegeburger with lentils, 74
Lentils. *See* Legumes or Vegetables.
Lettuce bread, 108

INDEX

Macaroni and bacon casserole, 92
Macaroni and tomato casserole, 64
Macaroni salad with tofu, 50
Mackeral crescent swirls, 165
Mackeral low cost fish cakes with TVP, 58
Mackeral sandwich spread with tofu, 50
Manicotti stuffed with tofu, 48
Meal planning with protein, 22
Measurements, 24–25
Meat and potato roll-ups with TVP, 64
Meatballs with tofu, 45
Meat, luncheon, 48
Meat Stretchers, 132–151
  Beef
    Barbecued beef roll-ups, 136
    Cassoulet, 148
    Corned beef hash (Peppers Parmigiana), 148
    Chuck, chili, 136
    Flank steak, Chinese style, 141
    Flank steak, Filipino style, 137
    Flank steak, tacos, 141
    Old fashioned vegetable soup, 149
    Planked ground beef, 134
    Teriyaki, 138
    Zucchini boats, 134
  Chicken
    Almond chicken, 145
    Chicken casserole, 146
    Chicken casserole, special, 146
    Chop suey, 142
    Pineapple, 143
    Sesame, 144
    Thighs, stuffed, 151
  Cooking tips, 134
  General information, 132
  Nutritive value, 132
  Pork
    Pork chops and cabbage casserole, 138
    Pork paprika, Filipino style, 140
    Pork with peas and bell peppers, 139
    Sweet and Sour pork, 139
  Turkey
    Corn Chowder, 150
    Ground turkey and tofu meatballs, 45
    Hot dogs, 73
    Parslied turkey tenderloin, 147
    Scallopini, 150
Mexican casserole with TVP, 62
Minestrone soup, soybean, 32
Muffins, coffee cake, Sunday morning, 100
Muffins, whole wheat, 104

Oatmeal patties with tofu, 43
Old fashioned vegetable soup, 149
Orange nut bread, 114

Pasta
  Calypso skillet with TVP, 59
  Cannelloni with TVP, 66
  Fettucine, 93
  Lasagne with tofu, 49
  Macaroni and bacon casserole, 92
  Macaroni salad with tofu, 50
  Manicotti stuffed with tofu, 48
  Soybean spaghetti sauce, 34
  Soybean spaghetti skillet, 34

INDEX

Tofu Sea Island salad, 52
Tomato macaroni casserole, 94
Pastry shell, 8″, 114
Pastry shell, 9″, 115
Peanut butter and honey snack balls, 94
Peas, blackeye and rice, 80
Peppers, green, stuffed with lentils, 72
Peppers Parmigiana, stuffed with corn beef hash, 148
Persimmon bread, 106
Pineapple cheese nut loaf, 108
Pinto beans, 78
Pinto beans (Portuguese style), 78
Poached fish, how to, 155
Pocket bread, 111
Pork
    Chops and cabbage casserole, 138
    Hash patties with tofu, 47
    Paprika pork, Filipino style, 140
    Patties with tofu, 48
    Pork or beef with tofu, 46
    Pork with peas and bell peppers, 139
    Spareribs with lentils, 74
    Sweet and sour, 139
Potato and meat roll-ups with TVP, 64
Potato pancakes (Puffer), 130
Protein, 9–22
Pumpkin bread, 106

Quiche Lorraine, 125
Quiche Lorraine, corn, 126

Refried beans, 78
Rice
    Rice, brown, mushroom stuffing, 164
    Rice casserole, 90
    Rice pilaf, Spanish style, 92
    Rice with blackeye peas, 80
Rolls, herb, with tofu, 52

Salads
    Chesapeake Bay fish salad, 157
    Fish salad or sandwich spread with tofu, 50
    Golden delight salad with soybeans, 30
    Macaroni salad with tofu, 50
    Sea Island salad with tofu, 52
    Soybean salad, 31
    Tabouli, 91
    Tuna and alfalfa sprouts salad, 57
    Yogurt and jello salad, 119
Sauces
    Barbecue sauce, 136
    Cannelloni sauce, 67
    Chile Rellenos sauce, 121
    Clam Marinara sauce, 61
    Dressing for Golden Delight salad with soybeans, 30
    Enchilada sauce, 60
    Light supreme sauce for chicken, 144
    Mushroom sauce for fish, 156
    Mustard sauce for gluten won tons, 89
    Seasame seed sauce for falafel, 76
    Soybean spaghetti sauce, 34
    Sweet and sour sauce for tofu, 46
    Teriyaki, 150
    Tomato and cheese sauce for lentils, 72
    Tomato sauce for eggplant, 128

INDEX

Tomato sauce for soybean patties, 37
White sauce for tofu fish croquettes, 44
Scallopini, 150
Scrambled eggs with tofu, 41
Sea Island salad with tofu, 52
Sesame chicken, 144
Sesame fish, 162
Shrimp chow mein, 160
Smoked fish with brown rice mushroom stuffing, 164
Sopaipillas (fried bread), 110
Soups
    Clam chowder, 158
    Old fashioned vegetable soup, 149
    Soybean gumbo soup, 32
    Soybean minestrone soup, 32
    Tofu cucumber soup, 42
    Tomato soup with tofu, 42
Soybeans, 26–38
    Baked, 36
    Chile, 33
    Cooking tips, 27
    Fresh green soybeans in butter, 30
    General information, 26
    Golden Delight salad and dressing, 30
    Gumbo soup, 32
    Minestrone soup, 32
    Nutritive values, 26
    Patties with tomato sauce, 37
    Soybean salad, 31
    Soy flakes, preparation of, 36
    Soy flakes timbales, 36
    Soy nuts, 38
    Spaghetti sauce, 34
    Spaghetti skillet, 34
    Tomato cheese casserole, 35
Spanish style baked fish, 160
Spanish style rice pilaf, 92
Spareribs with lentils, 74
Squash casserole, 130
Steak
    Flank, Chinese style, 141
    Flank, Filipino style, 137
    Flank, tacos, 141
Stuffed chicken thighs, 151
Stuffings
    Brown rice mushroom stuffing, 164
    Cornbread biscuit stuffing, 100
Sweet and sour broiled fish, 162
Sweet and sour pork, 139
Sweet and sour tofu, 46

Tabouli. *See* Bulgur.
Tacos with TVP, 63
Teriyaki, 138
Textured Vegetable Protein (TVP), 54–67
    Beef pie, ground, 65
    Calypso skillet, 59
    Cannelloni, 66
    Chicken or turkey curry, 56
    Clam Marinara sauce, 61
    Cooking tips, 55
    Curry, chicken or turkey, 56
    Enchiladas, 60
    Enchilada sauce, 60
    Fish cakes, low cost, 58
    General information, 54
    Meat and potato roll-ups, 64
    Mexican casserole, 62
    Nutritive value, 54
    Tacos, 63
    Tournament supreme, 66
    Turkey or chicken curry, 56
    Turkey patties, 57
Three bean casserole, 79

Three bean casserole with cheese, 79
Tofu, 39–53
  Beef or pork with tofu, 46
  Cabbage rolls, 51
  Coffee cake, 53
  Cooking tips, 40
  Cucumber soup, 42
  Fish croquettes, 44
    white sauce, 44
  Fish salad or sandwich spread, 50
  General information, 39
  Hamburgers, 44
  Lasagne, 49
  Macaroni salad, 50
  Manicotti stuffed with tofu, 48
  Meatballs, 45
  Nutritive value, 39
  Oatmeal patties, 43
  Patties, 48
  Pork or beef, 46
  Pork hash patties, 47
  Rolls, herb, 52
  Scrambled eggs, 41
  Sea Island salad, 52
  Sweet and sour, 46
  Tempura, 50
    batter, 51
  Tofu with vegetables, 43
  Tomato soup with tofu, 42
Tomato and cheese sauce with lentils, 72
Tomato cheese casserole with soybeans, 35
Tomato macaroni casserole, 94
Tomato soup with tofu, 42
Tortillas, 110
Tournament supreme with TVP, 66
Tuna and alfalfa sprouts salad, 157
Turkey
  Corn chowder, 150
  Curry with TVP, 56
  Hot dogs, 73
    See spiced baked lentils
  Patties with TVP, 57
  Scallopini, 150
  Tenderloin, parslied, 147

Vegeburger with lentils, 74
Vegetables
  Beans
    Dried beans and dried peas (any kind), 77
    Fava with tomato sauce, 77
    Garbanzo beans and chicken, 80
    Pinto beans, 78
    Pinto beans, Portuguese style, 78
    Refried beans, 78
    Three bean casserole, 79
    Three bean casserole with cheese, 79
  Cabbage rolls, stuffed with tofu, 51
  Chile Rellenos, 120
  Chile soybeans, 33
  Corn Quiche Lorraine, 126
  Egg Fu Yung, 124
  Eggplant Parmesan, 128
  Eggplant stuffed with cheese, chive and egg, 128
  Fish-vegetable bake, 163
  Gold and green casserole, 129
  Green chile and cheese casserole, 123
  Lentils, basic, 70
  Lentils, puree, 71
  Lentils, stewed, 71
  Lentils, spiced, baked, 73
  Lentil loaf with tomato and cheese sauce, 72
  Lentils vegeburger, 74
  Meat and potato roll-ups with TVP, 64

INDEX

Peas, blackeye and rice, 80
Peppers, green, stuffed with lentils, 72
Peppers Parmigiana, stuffed with corned beef, 148
Puffer (potato pancakes), 130
Soybeans, 26–38
   Baked, 36
   Chile, 33
   Fresh green soybeans in butter, 30
   Golden Delight salad and dressing, 30
   Gumbo soup, 32
   Minestrone soup, 32
   Soybean patties with tomato sauce, 37
   Soybean salad, 31
   Soy flakes, preparation of, 36
   Soy flakes timbales, 36
   Soy nuts, 38
   Spaghetti sauce with soybeans, 34
   Spaghetti skillet with soybeans, 34
   Tomato cheese casserole, 35

Squash casserole, 130
Tacos with TVP, 63
Three bean casserole, 79
Three bean casserole wih cheese, 79
Tofu with vegetables, 43
Tomato cheese casserole with soybeans, 35

Wheat, Boston baked, 86
Wheat. *See also* Bulgur and Gluten
Wheat germ banana nut bread, 109
Wheat, steamed, 85
Wheat, Tabouli. *See* Bulgur.
White fish, poached, 156

Yogurt banana dessert, 120
Yogurt fresh vegetable dip, 120
Yogurt, jello salad, 119
Yogurt pie with graham cracker crust, 119

Zucchini boats, 134
Zucchini gold and green casserole, 129